RECENT DEVELOPMENTS IN TIME-FREQUENCY ANALYSIS

edited by

Leon Cohen
Hunter College and Graduate Center of CUNY

Patrick Loughlin
University of Pittsburgh

A Special Issue of
MULTIDIMENSIONAL SYSTEMS AND SIGNAL PROCESSING
An International Journal

Volume 9, No. 4 (1998)

KLUWER ACADEMIC PUBLISHERS
Boston / Dordrecht / London

Distributors for North, Central and South America:
Kluwer Academic Publishers
101 Philip Drive
Assinippi Park
Norwell, Massachusetts 02061 USA
Telephone (781) 871-6600
Fax (781) 871-6528
E-Mail <kluwer@wkap.com>

Distributors for all other countries:
Kluwer Academic Publishers Group
Distribution Centre
Post Office Box 322
3300 AH Dordrecht, THE NETHERLANDS
Telephone 31 78 6392 392
Fax 31 78 6546 474
E-Mail <orderdept@wkap.nl>

 Electronic Services <http://www.wkap.nl>

Library of Congress Cataloging-in-Publication Data

A C.I.P. Catalogue record for this book is available
from the Library of Congress.

ISBN 978-1-4419-5065-9

Printed on acid-free paper.

Printed in the United States of America

MULTIDIMENSIONAL SYSTEMS AND SIGNAL PROCESSING

An International Journal

Volume 9, No. 4, October 1998

Special Issue: Recent Developments in Time-Frequency Analysis
Guest Editors: Leon Cohen and Patrick Loughlin

Multidimensional Systems and Signal Processing, 9, 337 (1998)
© 1998 Kluwer Academic Publishers, Boston. Manufactured in The Netherlands.

Editorial

This unique Special Issue grew out of a proposal to invite about a couple of dozen scientists to submit short, concise articles on recent developments in time-frequency analysis. The objective of the guest editors was to minimize the rehashing of old results and maximize the inclusion, in a direct clean, and concise manner, of those items that are considered to be important and new and likely to stimulate further research activities in the future. Both theory and application were targeted to receive equal emphasis. The uniqueness of the proposal demanded a special care in evaluation and Professor Marwan Simaan, a co-editor of this journal, and I became convinced of the beneficial effects of this type of issue especially with the popularity of the subject-matter and the experience of the guest coeditors, Professors Leon Cohen and Pat Loughlin. I hope that the readers will find the diversity of content and terse but clear and non-verbose style of exposition to be refreshing and helpful. I thank the guest coeditors for their efforts and Marwan Simaan for monitoring the progress towards its eventual submission to me. Thanks are also due to Melissa Fearon and Robert Holland of Kluwer Academic Publishers for bringing out this Issue in time, by carefully controlling the various stages of acquisition and production, for the meeting on Time-Frequency at Pittsburgh, during October 1998.

I am pleased to note that the change in the policy for direct submission of manuscripts to a centralized address supplied by Kluwer is working well and I feel that with Karen Cullen's cooperation and care, the prospective contributors as well as readers will be better served by prompt and fair reviews leading to quality publications. (Karen's position has now been filled by Melissa Sullivan and we look forward to the continuation of a pleasant and productive working atmosphere.) Finally, in the next Editorial I shall be able to report on the outcome of the First International Workshop on Multidimensional Systems at Zielona Gora, Poland, and the several special sessions devoted to the theory and applications of multidimensional systems scheduled at the Mathematical Theory of Networks and Systems (MTNS '98) Symposium at Padova, Italy.

N. K. Bose
Editor-in-Chief

Multidimensional Systems and Signal Processing, 9, 339 (1998)
© 1998 Kluwer Academic Publishers, Boston. Manufactured in The Netherlands.

Guest Editorial

Time-frequency analysis is a mature field which originated some fifty years ago when it was realized that many signals, both man-made and natural, have spectral content that changes in time. Examples include speech, many biological and biomedical signals, sonar, radar, communication signals, and waves propagating through dispersive media, among others. It therefore became necessary to develop methods that would reveal these spectral time variations and to correlate the variations with the physical mechanisms producing them. Since that time, many methods have been developed and become standard, and other time-varying signals have been uncovered and studied. In addition many new ideas have been introduced. Research in the area continues at a strong pace and the last ten years have been particularly active.

This volume of Multidimensional Systems and Signal Processing brings together a collection of 18 papers on recent developments in the field. Some of these include extensions to joint signal densities of arbitrary variables, new definitions of instantaneous frequency and AM-FM signal decompositions, methods for nonstationary signal classification and detection, applications in image processing, and new developments in Gabor and wave packet transforms. The papers in this issue present results in each of these areas, as well as other related topics in time-frequency analysis.

In order to accommodate a large number of authors and present a variety of results to readers, papers were deliberately kept short and focused. We hope that readers appreciate the variety of topics and that this issue stimulates further development in the field.

Finally, it is our pleasure to thank the reviewers whose comments and prompt and timely assistance were instrumental in compiling this issue, Marwan Simaan and Nirmal Bose for their encouragement and assistance in bringing about this special issue, and the production staff at Kluwer Academic Publishers for their patient help in putting it all together.

Leon Cohen
Pat Loughlin

Multidimensional Systems and Signal Processing, 9, 341–348 (1998)
© 1998 Kluwer Academic Publishers, Boston. Manufactured in The Netherlands.

Joint Distributions of Arbitrary Variables Made Easy

RICHARD G. BARANIUK*   richb@rice.edu
Department of Electrical and Computer Engineering, Rice University,   www.dsp.rice.edu
6100 South Main Street, Houston, TX 77005, USA

Received August 25, 1996; Accepted January 6, 1998

Abstract. We propose a simple framework for studying certain distributions of variables beyond time-frequency and time-scale. When applicable, our results turn the theory of joint distributions of arbitrary variables into an easy exercise of coordinate transformation. While straightforward, the method can generate many distributions previously attainable only by the general construction of Cohen, including distributions of time versus inverse frequency, time versus Mellin transform (scale), and time versus chirp. In addition to providing insight into these new signal analysis tools, warp-based distributions have efficient implementations for potential use in applications.

Key Words: time-frequency analysis, unitary equivalence

1. Introduction

The successful application of joint time-frequency distributions to problems in time-varying spectral analysis has motivated numerous studies into distributions of other variables for use when a strict time-frequency analysis is not appropriate [1].

Joint distributions generalise single variable distributions that measure the energy content of some physical quantity in a signal. Given a quantity a represented by the Hermitian (self-adjoint) operator $\mathcal{A}$, we obtain the density $|(\mathbb{F}_{\mathcal{A}}s)(a)|^2$ measuring the "a content" of the signal s simply by squaring the projection of s onto the formal eigenfunctions[1] $\mathbf{u}_a^{\mathcal{A}}$ of $\mathcal{A}$ [1]

$$|(\mathbb{F}_{\mathcal{A}}s)(a)|^2 \equiv \left| \int s(x) \left[\mathbf{u}_a^{\mathcal{A}}(x) \right]^* dx \right|^2. \tag{1}$$

Classical examples of single variable densities include the time density $|(\mathbb{F}_{\mathcal{T}}s)(t)|^2 = |s(t)|^2$ and the frequency density $|(\mathbb{F}_{\mathcal{F}}s)(f)|^2 = |S(f)|^2$ obtained by projecting onto the Dirac eigenfunctions of the time operator $(\mathcal{T}s)(x) \equiv x\,s(x)$ and the sinusoidal eigenfunctions of the frequency operator $(\mathcal{F}s)(x) \equiv \frac{1}{j2\pi}\dot{s}(x)$ [1]. (We will use both S and $\mathbb{F}s$ to denote the Fourier transform of the signal s.)

Joint distributions attempt to measure the simultaneous signal energy content of multiple physical quantities. Given two quantities a and b, a joint distribution $(\mathbf{P}_{a,b}s)(a,b)$ measuring the joint a-b content in the signal s has as marginals the respective $\mathcal{A}$ and $\mathcal{B}$ energy densities[2]

$$\int (\mathbf{P}_{a,b}s)(a,b)\,db = |(\mathbb{F}_{\mathcal{A}}s)(a)|^2, \tag{2}$$

---

*   This work was supported by the National Science Foundation, grant no. MIP–9457438, and by the Office of Naval Research, grant no. N00014–95–1–0849.

9

$$\int (\mathbf{P}_{a,b} s)(a, b) \, da = |(\mathbb{F}_B s)(b)|^2. \tag{3}$$

The Wigner distribution from Cohen's class of time-frequency distributions [1] supplies a classical example of a joint distribution that marginalises to the time and frequency densities.

Many different constructions have been proposed for generating joint distributions. The various approaches fall into two broad categories: *general methods* (difficult) and *coordinate change methods* (easy). General methods can create distributions for every possible pairing of physical quantities by working from first principles [1,5,6]. Coordinate change methods, on the contrary, sacrifice some flexibility for simplicity by bootstrapping existing distributions into new contexts using signal or axis transformations [5,7–10]

While not general purpose, coordinate change approaches to joint distributions remain attractive, because they provide a straightforward interpretation of the distributions they construct. Individually, each transformation method has severe limitations, however, as each technique can generate distributions for only a very restricted set of physical quantity pairs. The purpose of this paper is to demonstrate that a novel combination of two different coordinate change procedures can manufacture joint distributions for a much larger number of physical quantity pairs. In particular, our new procedure can generate most of the distributions considered as examples of the general method of Cohen in [1]. When applicable, our technique turns the difficult theory of joint distributions of arbitrary variables into an easy exercise of coordinate transformation.

2. Three Approaches to Joint Distributions

2.1. Cohen's General Method

Given two operators $\mathcal{A}$ and $\mathcal{B}$ representing two arbitrary physical quantities a and b, Cohen forms the joint distribution of a and b as (see [1,5] for more details)

$$(\mathbf{P}_{a,b} s)(a, b) = \iiint s^*(x) \left(e^{j2\pi(\alpha \mathcal{A} + \beta \mathcal{B})} s \right)(x) \, e^{-j2\pi(\alpha a + \beta b)} \, dx \, d\alpha \, d\beta. \tag{4}$$

Cohen's construction is general, but it requires that we solve a sometimes complicated operator equation to express the exponentiated operator $e^{j2\pi(\alpha \mathcal{A} + \beta \mathcal{B})}$. The time-frequency case is well understood; using the time and frequency operators $\mathcal{T}$ and $\mathcal{F}$ yields Cohen's class of time-frequency distributions [1, p. 136].

2.2. Axis Transformation Method

Joint a–b distributions are easily obtained when we can functionally relate these variables to time and frequency by $a = \alpha^{-1}(t)$ and $b = \beta^{-1}(f)$. In this special case, we can derive an a–b distribution simply by warping the coordinates of a time-frequency distribution $\mathbf{P}_{t,f}$ [5]

$$(\mathbf{P}_{a,b} s)(a, b) = |\dot{\alpha}(a)\,\dot{\beta}(b)| \; (\mathbf{P}_{t,f} s) [\alpha(a), \beta(b)]. \tag{5}$$

It is easily verified that all $\mathbf{P}_{a,b}s$ obtained in this way correctly marginalise to $|(\mathbb{F}_As)(a)|^2$ $= |\dot{\alpha}(a)|\,|(\mathbb{F}_Ts)(\alpha(a))|^2 = |\dot{\alpha}(a)|\,|s(\alpha(a))|^2$ and $|(\mathbb{F}_Bs)(b)|^2 = |\dot{\beta}(b)|\,|(\mathbb{F}_{\mathcal{F}}s)(\beta(b))|^2 = |\dot{\beta}(b)|\,|S(\beta(b))|^2$.

Example.   Distributions $\mathbf{P}_{t,r}s$ of time t and "inverse frequency" r (represented by the operator $\mathcal{R} = \frac{f_0}{\mathcal{F}}$) can be constructed from Cohen's class time-frequency distributions through the change of variable $r = \frac{f_0}{f}$ [1, p. 238]. The resulting class of distributions, obtained as

$$(\mathbf{P}_{t,r}\,s)(t,r) \;=\; \frac{f_0}{r^2}\,(\mathbf{P}_{t,f}\,s)\left(t,\frac{f_0}{r}\right), \tag{6}$$

marginalises to time $|s(t)|^2$ and inverse frequency $|(\mathbb{F}_{\mathcal{R}}s)(r)|^2 = \frac{f_0}{r^2}|S(f_0/r)|^2$.

2.3. Signal Transformation Method

Joint a–b distributions are also easily generated when the quantities a and b are *unitarily equivalent* to time and frequency [7,8], with

$$\mathcal{A} = \mathbf{U}^{-1}\mathcal{T}\mathbf{U}, \qquad \mathcal{B} = \mathbf{U}^{-1}\mathcal{F}\mathbf{U}, \tag{7}$$

and $\mathbf{U}$ a unitary transformation. In this case, a joint a–b distribution can be obtained by preprocessing a time-frequency distribution with $\mathbf{U}$

$$(\mathbf{P}_{a,b}\,s)(a,b) \;=\; (\mathbf{P}_{t,f}\,\mathbf{U}s)(a,b). \tag{8}$$

The signal transformation $\mathbf{U}$ can be interpreted as "rotating" the coordinates of the time-frequency distribution to the new variables.

All $\mathbf{P}_{a,b}s$ obtained in this way correctly marginalise to $|(\mathbb{F}_As)(a)|^2 = |(\mathbb{F}_T\mathbf{U}s)(a)|^2 = |(\mathbf{U}s)(a)|^2$ and $|(\mathbb{F}_Bs)(b)|^2 = |(\mathbb{F}_{\mathcal{F}}\mathbf{U}s)(b)|^2 = |(\mathbb{F}\mathbf{U}s)(b)|^2$ [7].

Example.   Define the logarithmic time operator $(\mathcal{L}s)(x) \equiv \log(x)\,s(x)$, and define the Mellin operator $\mathcal{H} \equiv \frac{1}{2}(\mathcal{T}\mathcal{F} + \mathcal{F}\mathcal{T})$. (Cohen refers to $\mathcal{H}$ as the "scale" operator [1,11].) These operators are unitarily equivalent to $\mathcal{T}$ and $\mathcal{F}$ as in (7) with $(\mathbf{U}s)(x) = e^{x/2}\,s(e^x)$. Therefore, we can construct distributions marginalising to logarithmic time $|(\mathbb{F}_{\mathcal{L}}s)(l)|^2 = |e^{l/2}\,s(e^l)|^2$ and Mellin transform

$$|(\mathbb{F}_{\mathcal{H}}s)(m)|^2 \;=\; \left|\int s(x)\,e^{-j2\pi m\log x}\,|x|^{-1/2}\,dx\,\right|^2 \tag{9}$$

simply by preprocessing the signal by $\mathbf{U}$ before computing a time-frequency distribution [7–10].

2.4. Shortcomings of Transformation Methods

While simple, both the axis and signal transformation methods place severe restrictions on the pairs of operators for which they can generate joint distributions. In the axis transformation method, we require the variables a and b to be warped versions of time and frequency.[3] In the signal transformation method, we require the operators $\mathcal{A}$ and $\mathcal{B}$ to be unitarily equivalent to time and frequency. For instance, in warping the frequency marginal to the Mellin marginal as above, we are obliged to accept logarithmic time for the other marginal.

3. Linking Signal and Axis Transformations

Due to the individual limitations of the axis and signal transformation methods, only general operator methods can generate joint distributions for all possible operator pairs. However, when used in *tandem*, the axis and signal transformation methods yield a powerful method for generating a large number of joint distributions. By executing first a unitary preprocessing transformation on the signal and then an axis warping transformation on the distribution, we can remap time-frequency distributions to a large class of different joint distributions.

THEOREM: *Let a and b be two variables whose Hermitian operator representations $\mathcal{A}$ and $\mathcal{B}$ satisfy the following two conditions: (i) $\mathcal{A}$ and $\mathcal{B}$ can be related to the time and frequency operators as*

$$\mathcal{A} = \mathbf{Z}^{-1} \mathcal{T} \mathbf{Z}, \qquad \mathcal{B} = \mathbf{U}^{-1} \mathcal{F} \mathbf{U}, \tag{10}$$

with $\mathbf{Z}$ and $\mathbf{U}$ unitary transformations. (ii) The composition $\mathbf{V} \equiv \mathbf{Z}\mathbf{U}^{-1}$ is an axis warping operator of the form

$$(\mathbf{V}g)(x) = |\dot{v}(x)|^{1/2} \, g[v(x)], \tag{11}$$

with v a smooth, 1–1 function. Then a joint a–b distribution $(\mathbf{P}_{a,b} s)(a, b)$ can be constructed from any time-frequency distribution $(\mathbf{P}_{t,f} s)(t, f)$ through

$$(\mathbf{P}_{a,b} s)(a, b) = |\dot{v}(a)| \, (\mathbf{P}_{t,f} \, \mathbf{U}s) \, [v(a), b]. \tag{12}$$

Proof: Direct evaluation of the marginals of the distribution (12) proves the result. For the a marginal, we have

$$\int (\mathbf{P}_{a,b} s)(a, b) \, db = |\dot{v}(a)| \int (\mathbf{P}_{t,f} \, \mathbf{U}s)[v(a), b] \, db$$

$$= |\dot{v}(a)| \, |(\mathbf{U}s)[v(a)]|^2$$

$$= |(\mathbf{V}\mathbf{U}s)(a)|^2 = |(\mathbf{Z}s)(a)|^2$$

$$= |(\mathbb{F}_\mathcal{A}s)(a)|^2. \tag{13}$$

For the b marginal, we have

$$\int (\mathbf{P}_{a,b} s)(a, b) \, da = \int (\mathbf{P}_{t,f} \, \mathbf{U}s)(v(a), b) \, |\dot{v}(a)| \, da$$

$$= \int (\mathbf{P}_{t,f} \, \mathbf{U}s)(x, b) \, dx$$

$$= |(\mathbb{F}_{\mathcal{F}} \, \mathbf{U}s)(b)|^2 = |(\mathbb{F}_{\mathcal{B}} s)(b)|^2. \tag{14}$$

∎

The interpretation of (12) is simple: The transformation $\mathbf{U}$ rotates the (t, f) coordinates of the time-frequency distribution to the new coordinates (u, b). The transformation $\mathbf{V}$ then warps the rotated time axis u to align it with the quantity a.

In other words, if $\mathcal{A}$ and $\mathcal{B}$ relate to $\mathcal{T}$ and $\mathcal{F}$ as in (10) and (11), then we can obtain a large class of a–b distributions using a simple three-step procedure: [4]

1. transform the signal $s \mapsto \mathbf{U}s$;

2. compute a conventional time-frequency distribution of the transformed signal;

3. warp the remapped time axis of the resulting distribution.

A numerical warping algorithm suitable for use on the digital computer is readily programmed.

The advantage of the double transformation method is that it breaks the severe restrictions placed on the quantities a and b by both the axis and signal transformation methods described in Section 2. By allowing the choice of both $\mathbf{U}$ and $\mathbf{Z}$, we gain access to a much larger class of distributions. However, completely free choice is still not possible, because $\mathbf{U}$ and $\mathbf{Z}$ must have the structural property of cancelling modulo the warping operator $\mathbf{V}$.

Several extensions to the theorem are straightforward. Reversing the rôles of time and frequency moves the warping transformation in (12) into the second argument. We also do not have to start from time-frequency distributions. We can map existing distributions based on arbitrary operators $\mathcal{C}, \mathcal{D}$ to distributions based on $\mathcal{A}, \mathcal{B}$ so long as all four operators obey relations analogous to (10). The advantage of starting from time-frequency distributions is simply that they are well understood and therefore aid in understanding the new a–b distributions.

4. Examples

Linked signal/axis transformations are especially useful for transforming time-frequency distributions to distributions of time versus a "warped frequency" variable. In this case, we set $\mathbf{Z}$ to the identity operator, choose $\mathbf{U}$ to be a warping operator based around the axis warping function v^{-1}, and set $\mathbf{V} = \mathbf{U}^{-1}$ to warp the rotated time axis back the time variable. The resulting distributions marginalise to time and the integral transform

$$\int (\mathbf{P}_{t,b} s)(t, b) \, dt = \left| \int s(x) \, e^{-j2\pi b v(x)} \, |\dot{v}(x)|^{1/2} \, dx \right|^2. \tag{15}$$

Each choice of v results in a different warped frequency transform matched to a distinct class of instantaneous frequency characteristics. The Fourier, Mellin, and chirp transforms result from the choices $v(x) = x$, $v(x) = \log x$, and $v(x) = |x|^c \, \text{sgn}(x)$, respectively.

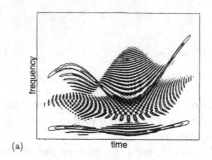 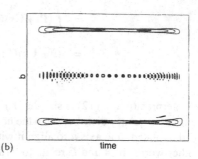

(a)   time   (b)   time

Figure 1. (a) Wigner time-frequency distribution of a synthetic test signal composed of two frequency modulated tones. (b) Warped Wigner distribution with time and "composite linear/sinusoidal instantaneous frequency" (variable "*b*" in (15)) marginals. The variable *b* measures the variation of the linear/sinusoidal instantaneous frequency path in time-frequency.

To continue the example of Section 2.3, applying the warp U^{-1} to the logarithmic time axis of the logarithmic time versus Mellin distribution remaps that axis back to true time. The resulting distributions lie in Cohen's class of time-Mellin distributions (time-scale in his terminology) [1,11]. This class contains the Marinovich-Altes (warped Wigner) distribution [9,12]. It is important to note that this class is out of the grasp of either signal or axis warping alone.

In Figure 1 we show two distributions of a signal consisting of two components concentrated along composite linear/sinusoidal instantaneous frequencies. Since the Wigner time-frequency distribution does not match signals of this type as well as sinusoids, impulses, and linear chirps, it exhibits copious cross-components. Prewarping the signal to account for the form of the signal yields a postwarped distribution that marginalises to time and "composite linear/sinusoidal instantaneous frequency" content (corresponding to variable "*b*" in (15)) and therefore better matches the signal.

Reversing the rôles of time and frequency in the warping procedure will yield frequency versus warped time distributions that match different classes of group delay (dispersion) characteristics.

5. Going Further

While our approach here has been aimed towards satisfying marginal distributions, analysis of the localization and covariance properties of warped time-frequency distributions is simple. An analysis similar to that in [7] shows that warped Cohen's class time-frequency distributions localize along warped impulses, warped sinusoids, and warped linear chirps. Unfortunately, except for the special case $Z = U$ in the theorem, most warped distributions lack any useful covariance to unitary transformations.[5] In fact, the time versus "warped frequency" distributions of Section 4 will not even be covariant to time shifts in general. While somewhat disappointing, we note that lack of covariance is a limitation of joint distributions of arbitrary variables in general and not just of this particular approach [13].

In certain situations we may desire simpler (yet non-orthogonal) functions of the form $e^{-j2\pi bv(x)}$ rather than $|\dot{v}(x)|^{1/2}\, e^{-j2\pi bv(x)}$ in the "warped frequency" integral transform on the right side of (15). Marginals of this type are not supported by the general construction of Section 2.1, since non-orthogonal eigenfunctions dictate non-Hermitian operators which in turn lead to complex-valued distributions. However, using non-unitary warping operators, we can construct distributions with these streamlined marginals by migrating the measure $|\dot{v}|$ into the left sides of the calculations (2), (3).

The signal/axis warping procedure can be applied to higher-order distributions such as the positive distributions of Cohen [1]. Positive distributions of arbitrary variables can also be set up directly, as

$$(\mathbf{P}^{pos}_{a,b}s)(a,b) \;=\; |(\mathbb{F}_{\mathcal{A}}s)(a)|^2\, |(\mathbb{F}_{\mathcal{B}}s)(b)|^2\, \Omega(a,b), \tag{16}$$

with Ω defined as in the time-frequency case [1, p. 198].

Acknowledgements

Thanks to Leon Cohen for suggesting this more in-depth analysis of warped time-frequency distributions. Thanks also to Faye Boudreaux-Bartels for valuable discussions and to Douglas Jones for suggesting the use of the JAM in this context.

Notes

1. When the operator $\mathcal{A}$ is Hermitian, the eigenequation is merely algebraic, and the eigenfunctions are actually tempered distributions. More rigorously, we could employ projection-valued measures for the eigenfunctions [2].
2. Alternatively, we can define joint distributions in terms of their covariance properties under certain unitary transformations. For more details, see [3,4].
3. In fact, it was precisely this shortcoming that lead Scully and Cohen to develop the general approach of Section 2.1.
4. If we define the 2-d transformation $\overline{\mathbf{V}}$ as the area-preserving change of variables

    $(\overline{\mathbf{V}}G)(x, y) \;\equiv\; |\dot{v}(x)|\, G\,[v(a), b]\,,$

    then we can write (12) in the standard form $\mathbf{P}_{a,b} = \overline{\mathbf{V}}\mathbf{P}_{t,f}\mathbf{U}$ found in [7]. However, whereas [7] emphasised using $\overline{\mathbf{V}}$ only to warp both axes of $\mathbf{P}_{a,b}s$ back to indicate time and frequency, in this paper, we exploit a range of different $\overline{\mathbf{V}}$.
5. By contrast, Cohen's class time-frequency distributions [1] are covariant to time and frequency shifts and affine class distributions [3,4] are covariant to time shifts and scale changes.

References

1. L. Cohen, *Time-Frequency Analysis*, Englewood Cliffs, NJ: Prentice-Hall, 1995.

2. G. B. Folland, *Harmonic Analysis in Phase Space*, Princeton, NJ: Princeton University Press, 1989.

3. O. Rioul and P. Flandrin, "Time-Scale Energy Distributions: A General Class Extending Wavelet Transforms," *IEEE Trans. Signal Processing*, vol. 40, 1992, pp. 1746–1757.

4. J. Bertrand and P. Bertrand, "A Class of Affine Wigner Functions with Extended Covariance Properties," *J. Math. Phys.*, vol. 33, 1992, pp. 2515–2527.

5. M. O. Scully and L. Cohen, "Quasi-Probability Distributions for Arbitrary Operators," in *Proceedings of nonlinear Dyamics and Chaos, Geometric Quantization and Wigner Functions* (Y. S. Kim and W. W. Zachary, eds.), (Berlin), Springer-Verlag, pp. 253–263, 1987.

6. R. G. Baraniuk, "Beyond Time-Frequency Analysis: Energy Densities in One and Many Dimensions," *IEEE Trans. Signals Processing*, vol. 46, no. 9, 1998.

7. R. G. Baraniuk and D. L. Jones, "Unitary Equivalence: A New Twist on Signal Processing," *IEEE Trans. Signal Processing*, vol. 43, 1995, pp. 2269–2282.

8. R. G. Baraniuk and L. Cohen, "On Joint Distributions of Arbitrary Variables," *IEEE Signal Processing Letters*, vol. 2, 1995, pp. 10–12.

9. R. A. Altes, "Wideband, Proportional-Bandwidth Wigner-Ville Analysis," *IEEE Trans. Acoust., Speech, Signal Processing*, vol. 38, 1990, pp. 1005–1012.

10. A. Papandreou, F. Hlawatsch, and G. F. Bourdreaux-Bartels, "The Hyperbolic Class of Quadratic Time-Frequency Representations. Part 1: Constant-Q Warping, the Hyperbolic Paradigm, Properties, and Members," *IEEE Trans. Signal Processing*, vol. 41, 1993, pp. 3425–3444.

11. L. Cohen, "The Scale Representation," *IEEE Trans. Signal Processing*, vol. 41, 1993, pp. 3275–3292.

12. G. Eichmann and N. M. Marinovich, "Scale-Invariant Wigner Distribution," in *Proc. SPIE Int. Soc. Opt. Eng.*, vol. 519, 1984, pp. 18–25.

13. R. G. Baraniuk, "Covariant Time-Frequency Representations Through Unitary Equivalence," *IEEE Signal Processing Letters*, vol. 3, 1996, pp. 79–81.

Multidimensional Systems and Signal Processing, 9, 349–354 (1998)
© 1998 Kluwer Academic Publishers, Boston. Manufactured in The Netherlands.

On the Use of Spatial Time Frequency Distributions for Signal Extraction

ADEL BELOUCHRANI   adel@ece.vill.edu
Department of Electrical and Computer Engineering, Villanova University, Villanova PA 19085

MOENESS G. AMIN*   moeness@ece.vill.edu
Department of Electrical and Computer Engineering, Villanova University, Villanova PA 19085

Received August 29, 1996; Revised March 25, 1997

Abstract. This paper deals with the extraction of signals from their instantaneous linear mixtures using time-frequency distributions. Fundamentally, this problem is a signal synthesis from the time-frequency (t-f) plane. However with the incorporation of the spatial information provided by a multisensor array, the problem can be posed as special case of blind source separation. So far, the blind source separation has been solved using only statistical information available on the source signals. Herein, we propose to solve the aforementioned problem using time-frequency signal representations and the spatial array aperture. The proposed approach relies on the difference in the t-f signatures of the sources to be separated. It is based on the diagonalization of a combined set of spatial time-frequency distribution matrices. A numerical example is provided to illustrate the effectiveness of our method.

Key Words: time-frequency distribution, spatial signal processing, blind signal estimation

1. Introduction

In statistical signal and array processing, the problem of signal extraction has received the name of blind source separation, which becomes an emerging field of fundamental research with a broad range of applications. It is motivated by practical problems that involve several source signals and several sensors. Each sensor receives a linear mixture of the source signals. The problem of the blind source separation consists then of recovering the original waveforms of the sources without any knowledge of the mixture structure. This mixture is often a convolutive mixture. However, in this paper we focus on the blind identification of an instantaneous linear mixture, which corresponds to a linear memoryless channel. This choice is motivated not only by the fact that such model is mathematically tractable, but also by the applicability to various areas, including semiconductor manufacturing process [1], factor analysis [2], narrowband signal processing [3], and image reconstruction [4].

So far, the problem of blind source separations has been solved using statistical information available on the source signals. The first solution to the source separation problem was proposed almost a decade ago [5] and was based on the cancellation of higher order moments assuming non-Gaussian and i.i.d source signals. Other criteria based on minimizations of cost functions, such as the sum of square fourth order cumulants [6], contrast functions [6] or likelihood function [7], have been used by several researchers. In the case of non i.i.d source signals and even Gaussian sources, solutions based on second order statistics are possible [8].

* This work is supported by the Rome Lab., contract # F30602-96-C-0077

17

When the frequency content of the source signals is time-varying, one can take advantage of the powerful tool of time-frequency distributions to separate and recover the incoming signals. The underlying blind source separation problem can be posed as a signal synthesis [9] from the t-f plane with the incorporation of the spatial diversity provided by the multisensor array. In this case, no masking is required and cross-terms no longer represents ambiguity in the synthesis of individual components. Herein, we introduce a new blind identification technique based on a joint diagonalization of a combined set of spatial time-frequency distribution matrices. The new approach exploits the difference between the t-f signatures of the sources. This method presents a number of attractive features. In contrast to blind source separation approaches using second-order and/or high order statistics, the proposed approach allows separation of Gaussian sources with identical spectra. Moreover, the effects of spreading the noise power while localizing the source energy in the time-frequency domain amounts to increasing the signal to noise ratio (SNR).

2. Data Model

Consider m sensors receiving an instantaneous linear mixture of signals emitted from n sources. The $m \times 1$ vector $\mathbf{x}(t)$ denotes the output of the sensors at time instant t which may be corrupted by an additive noise $\mathbf{n}(t)$. Hence, the linear data model is given by:

$$\mathbf{x}(t) = \mathbf{A}\mathbf{s}(t) + \mathbf{n}(t), \tag{1}$$

where the $m \times n$ matrix $\mathbf{A}$ is called the 'mixing matrix'. The n source signals are collected in the $n \times 1$ vector $\mathbf{s}(t)$. The mixing matrix $\mathbf{A}$ is full column rank but is otherwise unknown. In contrast with traditional parametric methods, no specific structure of the mixture matrix is assumed. The problem of blind source separation has two inherent indeterminacies such that the source signals can only be identified up to a fixed permutation and some complex factors [8]. In the underlying problem, we are concerned with the separation of signals which may have similar spectra, but still possess different structures and localization properties in the time-frequency domain.

3. Spatial Time-Frequency Distributions

The Cohen's class of time-frequency distributions (TFD) [10] of the signal $x(t)$ is given by

$$D_{xx}(t, f) = \int_{-\infty}^{\infty} \int_{-\infty}^{\infty} \phi(t - u, \tau) x(u + \tau/2) x^*(u - \tau/2) e^{-j2\pi f \tau} du d\tau \tag{2}$$

where t and f represent the time index and the frequency index, respectively. The kernel $\phi(t, \tau)$ is a function of the time and lag variables. The cross-TFD of two signals $x_1(t)$ and $x_2(t)$ is defined by

$$D_{x_1 x_2}(t, f) = \int_{-\infty}^{\infty} \int_{-\infty}^{\infty} \phi(t - u, \tau) x_1(u + \tau/2) x_2^*(u - \tau/2) e^{-j2\pi f \tau} du d\tau \tag{3}$$

Expressions (2) and (3) are now used to define the following data *spatial time-frequency distribution* (STFD) *matrix,*

$$\mathbf{D_{XX}}(t, f) = \int_{-\infty}^{\infty} \int_{-\infty}^{\infty} \phi(t - u, \tau)\mathbf{x}(u + \tau/2)\mathbf{x}^*(u - \tau/2)e^{-j2\pi f\tau}dud\tau \qquad (4)$$

where $[\mathbf{D_{XX}}(t, f)]_{ij} = D_{x_i x_j}(t, f)$, for $i, j = 1, \dots, n$.

A more general definition of the spatial time-frequency distribution matrix is given by,

$$\mathbf{D_{XX}}(t, f) = \int_{-\infty}^{\infty} \int_{-\infty}^{\infty} \Phi(t - u, \tau) \odot \mathbf{x}(u + \tau/2)\mathbf{x}^*(u - \tau/2)e^{-j2\pi f\tau}dud\tau \qquad (5)$$

where $\odot$ designs the Hadamard product, and $[\Phi(t, \tau)]_{ij} = \phi_{ij}(t, \tau)$ is the kernel associated with the pair of the sensor data $x_i(t)$ and $x_j(t)$.

Under the assumption of the linear data model of Eq. (1) and neglecting the noise, the STFD matrix takes the following simple structure:

$$\mathbf{D_{XX}}(t, f) = \mathbf{A}\mathbf{D_{SS}}(t, f)\mathbf{A}^H \qquad (6)$$

where the superscript ^{H} denotes the complex conjugate transpose of a matrix and $\mathbf{D_{SS}}(t, f)$ are the signal TFD matrices. We note that $\mathbf{D_{XX}}(t, f)$ is of dimension $m \times m$, whereas $\mathbf{D_{SS}}(t, f)$ is of $n \times n$ dimension. For narrowband array signal processing applications, matrix $\mathbf{A}$ holds the spatial information, and through a similarity transformation, it maps the auto- and cross-TFDs of the sources into auto- and cross-TFDs of the data.

Expression (6) is similar to that which commonly used in blind source separation [8] and direction of arrival (DOA) estimation problems, relating the signal correlation matrix to the data spatial correlation matrix. The two subspaces spanned by the principle eigenvectors of $\mathbf{D_{XX}}(t, f)$ and the columns of $\mathbf{A}$ are, therefore, identical. Since the off-diagonal elements are cross-terms of $\mathbf{D_{SS}}(t, f)$, then this matrix is diagonal for all (t-f) points which correspond only to the signal auto-terms. In the sequel, we consider the (t-f) points, which verify such property. In practice, to simplify the selection of such points of high power localization, we apply the smoothing kernel $\phi(t, \tau)$ that significantly decreases the contribution of the cross-terms in the t-f plane.

4. A Time Frequency Separation Principle

Let $\mathbf{W}$ denote a $m \times n$ matrix, such that $(\mathbf{WA})(\mathbf{WA})^H = \mathbf{UU}^H = \mathbf{I}$, i.e. $\mathbf{WA}$ is a $m \times m$ unitary matrix (this matrix is referred to as a whitening matrix, since it whitens the signal part of the observations and can be obtained form the eigendecomposition of the autocorrelation matrix of the data $\mathbf{x}(t)$, see [8] for more details). Pre- and post-multiplying the STFD-matrices $\mathbf{D_{XX}}(t, f)$ by $\mathbf{W}$, we then define the *whitened STFD-matrices* as:

$$\underline{\mathbf{D}}_{\mathbf{XX}}(t, f) = \mathbf{W}\mathbf{D_{XX}}(t, f)\mathbf{W}^H \qquad (7)$$

From the definition of $\mathbf{W}$ and Eq. (6), we may express $\underline{\mathbf{D}}_{\mathbf{XX}}(t, f)$ as

$$\underline{\mathbf{D}}_{\mathbf{XX}}(t, f) = \mathbf{U}\mathbf{D_{SS}}(t, f)\mathbf{U}^H \qquad (8)$$

Since the matrix $\mathbf{U}$ is unitary and $\mathbf{D_{SS}}(t, f)$ is diagonal, expression (8) shows that any whitened data STFD-matrix is diagonal in the basis of the columns of the matrix $\mathbf{U}$ (the eigenvalues of $\underline{\mathbf{D}}_{\mathbf{xx}}(t, f)$ being the diagonal entries of $\mathbf{D}_{\mathbf{SS}}(t, f)$).

If, for a (t-f) point, the diagonal elements of $\mathbf{D}_{\mathbf{SS}}(t, f)$ are all distinct, the missing unitary matrix $\mathbf{U}$ may be 'uniquely' (i.e. up to permutation and phase shifts) retrieved by computing the eigendecomposition of $\underline{\mathbf{D}}_{\mathbf{xx}}(t, f)$. However, when the t-f signatures of the different signals are not highly overlapping or frequently intersecting, which is likely to be the case, the selected (t-f) point often corresponds to one signal auto-term, rendering matrix $\mathbf{D}_{\mathbf{SS}}(t, f)$ deffient. That is, only one diagonal element of $\mathbf{D}_{\mathbf{SS}}(t, f)$ is different from zero. It follows that the determination of the matrix $\mathbf{U}$ from the eigendecomposition of a single whitened data STFD-matrix is no longer 'unique' in the sense defined above.

The situation is more favorable when considering *joint diagonalization* of a combined set $\{\underline{\mathbf{D}}_{\mathbf{xx}}(t_i, f_i) \mid i = 1, \ldots, p\}$ of p STFD matrices. This amounts to incorporating several (t-f) points in the source separation problem. It is noteworthy that two source signals with identical t-f signatures can not be separated even with the inclusion of all information in the t-f plane.

The *joint diagonalization* can be explained by first noting that the problem of the diagonalization of a single $n \times n$ normal matrix $\mathbf{M}$ is equivalent to the minimization of the criterion $C(\mathbf{M}, \mathbf{V}) \overset{\text{def}}{=} -\sum_i |\mathbf{v}_i^* \mathbf{M} \mathbf{v}_i|^2$ over the set of unitary matrices $\mathbf{V} = [\mathbf{v}_1, \ldots, \mathbf{v}_n]$ [11]. Hence, the Joint diagonalization of a set $\{\mathbf{M}_k \mid k = 1 \ldots p\}$ of p arbitrary $n \times n$ matrices is defined as the minimization of the criterion:

$$C(\mathbf{V}) \overset{\text{def}}{=} -\sum_k C(\mathbf{M}_k, \mathbf{V}) = -\sum_{ki} |\mathbf{v}_i^* \mathbf{M}_k \mathbf{v}_i|^2 \tag{9}$$

under the same unitary constraint. An efficient algorithm for solving (9) already exists in [8] and is the generalization of the Jacobi technique [11].

5. An Example

Here we present an illustration that involves real data signals. Two speech signals sampled at 8000 Hz are mixed by the following mixing matrix,

$$\mathbf{A} = \begin{bmatrix} 1.0 & 0.5 \\ 0.6 & 1.0 \\ 0.4 & 0.8 \end{bmatrix}.$$

The kernel used for the computation of the TFDs is the Choi-Williams kernel [10], which provides a good cancellation of the cross-terms. Four STFD matrices are considered. The corresponding (t-f) points are those of the highest power in the t-f plane. The TFDs of the two individual speech signals are shown in Fig. 1. Speech 1 and 2 of a male speaker are the words "Cars" and "Cats", respectively. The TFDs of the observed speech signals at three sensors are displayed in Fig. 2. Figure 3 shows the TFDs of the estimated speech signals by the proposed approach.

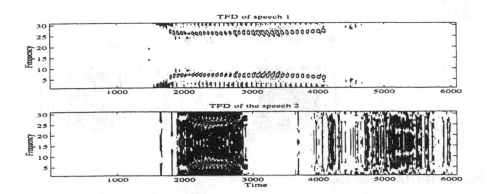

Figure 1. TFDs of the individual speeches.

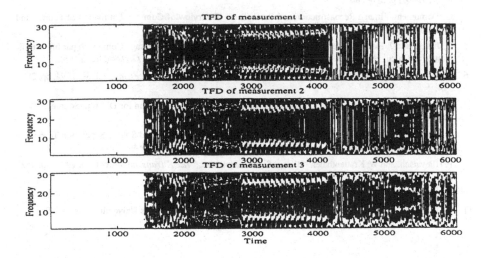

Figure 2. TFDs of the measurements.

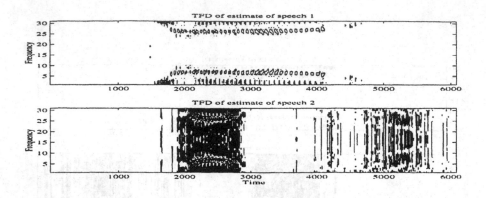

Figure 3. TFDs of the estimate signals.

References

1. C. M. Berrah, "Parameter Yield Estimation for a MOSFET Integrated Circuit," *Proc. 1990 IEEE ISCAS*, 1990, pp. 2260–2263.

2. E. E. Cureton, and R. B. D'Agostino, *FACTOR ANALYSIS An Applied Approach*, Lawrence Erlbaum Associates, 1983.

3. R. Schmidt, "Multiple Emitter Location and Signal Parameter Estimation," *IEEE Trans. on AP*, vol. 34, no. 1, 1986, pp. 276–280.

4. G. Demoment, "Image Reconstruction and Restoration: Overview of Common Estimation Structures and Problems," *IEEE Trans. on ASSP*, vol. 37, 1989, pp. 2024–22036.

5. C. Jutten, and J. Hérault, "Détection de Grandeurs Primitives Dans un Message Composite par une Architecture de Calcul Neuromimétrique en Apprentissage non Supervisé," *Proc. Gretsi*, (Nice), 1985.

6. P. Comon, "Independent Compotent Analysis, A New Concept?," *Signal Processing*, vol. 36, 1994, pp. 287–314.

7. A. Belouchrani, and J.-F. Cardoso, "Maximum Likelihood Source Separation for Discrete Sources," *Proc. EUSIPCO*, 1994, pp. 768–771.

8. A. Belouchrani, K. Abed Meriam, J.-F. Cardoso, and E. Moulines, "A Blind Source Separation Technique Using Second Order Statistics," *IEEE Trans. on SP*, vol. 45, 1997, pp. 434–444.

9. F. Hlawatsch, and W. Krattenthaler, "Bilinear Signals Synthesis," *IEEE Trans. on SP*, vol. 40, 1992, pp. 352–363.

10. L. Cohen, *Time-Frequency Analysis*, Prentice Hall, 1995.

11. G. H. Golub, and C. F. Van Loan, *Matrix Computations*, The Johns Hopkins University Press, 1989.

Multimedia Systems and Signal Processing, 9, 355–362 (1998)

On the Statistics of Spectrogram Reassignment Vectors[*]

ERIC CHASSANDE-MOTTIN   echassan@physique.ens-lyon.fr

PATRICK FLANDRIN   flandrin@physique.ens-lyon.fr
*Laboratoire de Physique, Ecole Normale Supérieure de Lyon, 46 allée d'Italie,
69364 Lyon Cedex 07, France*

FRANÇOIS AUGER   auger@ge44.univ-nantes.fr
GE44, CRTT, Bd de l'Université, BP 406, 44602 Saint-Nazaire Cedex, France

Received August 20, 1996; Revised February 8, 1997

Abstract. Reassignment is a non-linear technique which can improve on the localization of a spectrogram by moving its values according to a suitable vector field. Statistical properties of spectrogram reassignment vectors are investigated in detail. Closed form expressions are given when the observation consists in a non-random component embedded in white Gaussian circular noise, and when the analysis window is Gaussian. An extension to arbitrary windows is also proposed and theoretical claims are supported by numerical simulations.

Keywords: time-frequency distributions, reassignment methods

1. Introduction

Reassignment is a non-linear technique which has recently been reintroduced and put forward as an efficient means of getting sharply localized time-frequency distributions [1]. The technique consists in moving the value of a time-frequency distribution from the point where it has been computed to a new location which is more representative of the local signal energy distribution. Reassignment is based on a vector field which conveys a lot of information about the signal structure and it is therefore an important issue to quantify properties of this vector field, especially from a statistical point of view. This paper is devoted to first theoretical investigations of this question which — up to now — has received no attention in the literature, although it is believed to be of a key importance for many further signal processing applications based on reassigned distributions. While reassignment is a very general principle which can be applied to almost any time-frequency distribution, discussion will be restricted here to spectrograms only. Let us finally note that, due to space limitations, most proofs of the results presented here have been omitted, but they can be found in [2].

2. Basics and assumptions

In order to reassign the spectrogram associated to the short-time Fourier transform (STFT)

---

[*] This work was supported in part by the GdR 134 CNRS "Traitement du Signal et Images."

$$F_x^{(h)}(t, \nu) = \int_{-\infty}^{+\infty} x(t - \tau)h^*(\tau)e^{2\pi i\nu\tau} \, d\tau \tag{1}$$

based on a window $h(t)$, it is necessary to introduce two auxiliary STFT's based on the windows $(\mathcal{T}h)(t) = t\, h(t)$ and $(\mathcal{D}h)(t) = (d/dt)h(t)$. Given these three STFT's, reassignment amounts to shift the value of $\left| F_x^{(h)}(t, \nu) \right|^2$ from (t, ν) to a new location $(\hat{t}(t, \nu), \hat{\nu}(t, \nu))$ characterized by [1]

$$\left(\hat{t}(t, \nu),\, \hat{\nu}(t, \nu) \right) = \left(t - \mathrm{Re}\left\{ \frac{F_x^{(\mathcal{T}h)}(t, \nu)}{F_x^{(h)}(t, \nu)} \right\},\, \nu + \frac{1}{2\pi}\mathrm{Im}\left\{ \frac{F_x^{(\mathcal{D}h)}(t, \nu)}{F_x^{(h)}(t, \nu)} \right\} \right). \tag{2}$$

Identifying the time-frequency plane with the complex plane, it is therefore possible to define at each time-frequency point a dimensionless reassignment vector $r(t, \nu)$ according to

$$r(t, \nu) = \frac{\hat{t}(t, \nu) - t}{\Delta t_h} + i\frac{\hat{\nu}(t, \nu) - \nu}{\Delta\nu_h}, \tag{3}$$

where Δt_h and $\Delta\nu_h$ stand, respectively, for the mean-square duration and bandwidth of the analysis window. This study is devoted to statistical properties of $r(t, \nu)$ in the case of observations corrupted by a zero-mean noise $n(t)$ assumed to be white, Gaussian and analytic (hence, circular [3]), i.e., such that

$$\mathbf{E}[\mathrm{Re}\{n(t)\}\mathrm{Re}\{n(s)\}] = \mathbf{E}[\mathrm{Im}\{n(t)\}\mathrm{Im}\{n(s)\}] = \frac{\sigma^2}{2}\delta(t - s);\ \mathbf{E}[n(t)n(s)] = 0 \tag{4}$$

for any $(t, s) \in \mathbf{R}^2$, with its real and imaginary parts a Hilbert transform pair.

3. Statistics of reassignment vectors for Gabor spectrograms

Let us first consider a Gaussian window $h(t) = 2^{1/4}\lambda^{-1/2}\exp(-\pi(t/\lambda)^2)$, a situation referred to as a Gabor spectrogram. In this case, we have $\Delta t_h = (4\pi)^{-1/2}\lambda$ and $\Delta\nu_h = (4\pi)^{-1/2}/\lambda$, and $(\mathcal{T}h)(t) = t\, h(t)$ and $(\mathcal{D}h)(t) = (d/dt)h(t)$ happen to be proportional to each other. The relative reassignment vector reduces to

$$r(t, \nu) = -\frac{\sqrt{4\pi}}{\lambda}\frac{F_2}{F_1}, \tag{5}$$

where F_1 and F_2 are simplified notations for $F_x^{(h)}(t, \nu)$ and $F_x^{(\mathcal{T}h)}(t, \nu)$, respectively.

3.1. The "noise only" case

In the "noise only" case where the observation is $x(t) = n(t)$, it follows from properties related to both linear filtering and analytic circularity that the vector $\underline{F} = [F_1\, F_2]^t$ is zero-mean and Gaussian circular, with a joint probability density function (pdf) given by

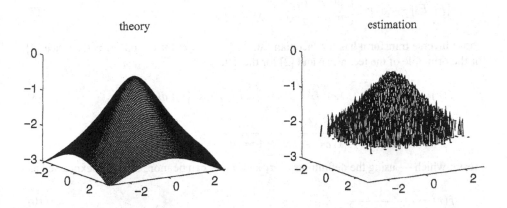

Figure 1. Log-probability density function of reassignment vectors in the Gabor spectrogram case and in a "noise only" situation. Left: theoretical joint density, as predicted by eq.(10). Right: estimated joint density, obtained from a simulation based on an average over 3 noise realizations, each of 412×156 time-frequency data points.

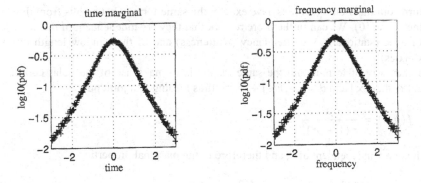

Figure 2. Log-marginals of the probability density function of reassignment vectors in the Gabor spectrogram case and in a "noise only" situation. Left: time marginal, right: frequency marginal. Solid lines correspond to the theoretical predictions of eq.(11), whereas the crosses have been obtained from a simulation based on an average over 3 noise realizations, each of 412×156 time-frequency data points.

$$f_{\underline{F}}(F_1, F_2) = \frac{1}{\pi^2 |\det \mathbf{\Gamma}|} \exp(-\underline{F}^\dagger \mathbf{\Gamma}^{-1} \underline{F}); \quad \mathbf{\Gamma} = \begin{bmatrix} \sigma_1^2 = 2\sigma^2 & 0 \\ 0 & \sigma_2^2 = \sigma^2 \lambda^2 / 2\pi \end{bmatrix}. \quad (6)$$

Making then the change of variables

$$[F_1 \ F_2] \rightarrow [w \ r] \ \begin{cases} w = F_1 \\ r = -(\sqrt{4\pi}/\lambda) \, F_2/F_1 \end{cases} \tag{7}$$

whose inverse transform has for Jacobian $|\lambda w|^2/4\pi$, we get for the pdf of $r(t, \nu)$ (see [4] for the principle of the technique and [2] for the details):

$$f(r) = \frac{\lambda^2}{4\pi} \iint_{-\infty}^{+\infty} |w|^2 f_{\underline{F}} \left(w, -\frac{\lambda}{\sqrt{4\pi}} r w \right) d\mathrm{Re}\{w\} \, d\mathrm{Im}\{w\} \tag{8}$$

$$= \frac{\lambda^2}{4\pi^2(\sigma_1^2 \sigma_2^2) \left[1/\sigma_1^2 + |\lambda r|^2/(4\pi\sigma_2^2) \right]^2}, \tag{9}$$

a result which — using the definitions of σ_1 and σ_2 — can be more simply reexpressed as

$$f(r) = \frac{1}{\pi \left(1 + |r|^2\right)^2}. \tag{10}$$

As a consequence, $r(t, \nu)$ appears as being zero-mean but of infinite variance, with marginal densities given by (let us recall that the real and imaginary parts of r correspond to the reassignment components in time and frequency, respectively)

$$f(\mathrm{Re}\{r\}) = \frac{1}{2(1 + (\mathrm{Re}\{r\})^2)^{3/2}}; \qquad f(\mathrm{Im}\{r\}) = \frac{1}{2(1 + (\mathrm{Im}\{r\})^2)^{3/2}}. \tag{11}$$

It turns out that both quantities have exactly the same form, what results from the radial symmetry of (10). We can furthermore remark that they are independent of the noise level σ^2, of time (stationarity), of frequency (whiteness) and of the window length (reduced coordinates).

Another way of formulating the same results is to make use of the polar coordinates $\rho = |r| \in [0, +\infty[$ and $\theta = \arg r \in] - \pi, \pi]$, thus leading to a joint pdf

$$f(\rho, \theta) = \frac{\rho}{\pi \left(1 + \rho^2\right)^2} \tag{12}$$

which is now independent of θ, and therefore to the marginal properties

$$f(\rho) = 2\rho / \left(1 + \rho^2\right)^2; \qquad f(\theta) = 1/2\pi. \tag{13}$$

3.2. The "signal + noise" case

The previous analysis can be carried over to "signal + noise" situations where the observation is of the form $x(t) = s(t) + n(t)$, with $s(t)$ a non-random component and $n(t)$ as in (4). According to this model, $x(t)$ is still Gaussian circular, although not zero-mean.

Calculations are in this case more cumbersome than in the "noise only" case but the derivation proceeds along the same lines (details are given in [2]). The final result is

$$f(r) = \frac{1}{\pi \left(1 + |r|^2\right)^2} \left[1 + \frac{S}{2\sigma^2} \frac{|1 + rr_0^*|^2}{1 + |r|^2} \right] \exp \left\{ -\frac{S}{2\sigma^2} \frac{|r - r_0|^2}{1 + |r|^2} \right\}, \tag{14}$$

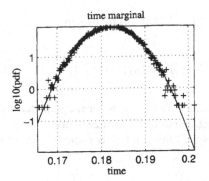
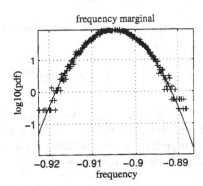

Figure 3. Log-marginals of the probability density function of reassignment vectors in the Gabor spectrogram case and in a "chirp+noise" situation (SNR = 64dB). Left: time marginal, right: frequency marginal. Solid lines correspond to the theoretical predictions derived from eq.(14), whereas the crosses have been obtained from a simulation based on an ensemble average over 20, 000 noise realizations.

with $S = \left| F_s^{(h)}(t, \nu) \right|^2$ and r_0 the noise-free reassignment vector.

The pdf now becomes a function of the local signal-to-noise ratio SNR = $S/2\sigma^2$. In the case where SNR $\rightarrow$ 0, one can check that (14) reduces to the "noise only" pdf (10). One can also remark that (14) tends to be approximatively Gaussian for small r's and, if one has furthermore SNR $\rightarrow +\infty$ ("signal only" case), one gets $f(r) \sim$ SNR exp $(-$SNR$) \sim$ 0 unless $r = r_0$, in which case $f(r) \sim$ SNR $\sim +\infty$.

The theoretical results of this Section are well-supported by numerical simulations, as illustrated in Figures 1 to 3.

4.   Extension to spectrograms with arbitrary windows

The use of Gaussian windows only may appear as a restriction and it is desirable to consider more general situations. Dealing with arbitrary windows requires however a specific treatment which cannot take advantage of the simplifications which occurred in the Gaussian case. In fact, it turns out that the problem of evaluating the pdf of a re-assignment vector based on an arbitrary window can be embedded in a more general problem which can be stated as follows: given an observation $x(t) = s(t) + n(t)$, where $s(t)$ is non-random and $n(t)$ as in (4), what is the joint pdf of the $(N - 1)$-dimensional complex-valued vector $\underline{r} = [r_1 \, r_2 .. r_{N-1}]^t$ whose coordinates are given by $(r_n = (1/\lambda_n)y_{n+1}/y_1; n = 1 \ldots N - 1)$, where the λ_n's are complex-valued numbers and the collection of y_n's results from the linear filtering of $x(t)$ by a family of filters of impulse responses $h_n(t)$?

This problem can be solved by introducing the change of variables

$$[y_1\, y_2 \cdots y_N] \longrightarrow [y_1\, r_1 \cdots r_{N-1}], \tag{15}$$

whose inverse transform has for Jacobian $\left| r_1^{N-1} \prod_{n=1}^{N-1} \lambda_n \right|^2$. It follows that

$$f(\underline{r}) = \left| \prod_{n=1}^{N-1} \lambda_n \right|^2 \iint_{-\infty}^{+\infty} |y_1|^{2(N-1)} f_{\underline{Y}}(\underline{r}_\lambda y_1)\, d\mathrm{Re}\{y_1\}\, d\mathrm{Im}\{y_1\}, \tag{16}$$

with $\underline{r}_\lambda = [1\ \lambda_1 r_1\ \lambda_2 r_2 \ldots \lambda_{N-1} r_{N-1}]^t$ and where $f_{\underline{Y}}(.)$ stands for the joint density of the filtered observations $(y_1, y_2, \ldots y_N)$, a Gaussian pdf of covariance matrix Γ.

This expression can be evaluated in a more explicit form and, after some tedious manipulations (for details, see [2]), we get

$$f(\underline{r}) = \frac{1}{\pi^N |\det \Gamma|} \left| \prod_{n=1}^{N-1} \lambda_n \right|^2 e^{-|b|^2/a+c} \\ \iint_{-\infty}^{+\infty} \left| y_1 + \frac{b}{a} \right|^{2(N-1)} \exp\left(-a|y_1|^2\right)\, d\mathrm{Re}\{y_1\}\, d\mathrm{Im}\{y_1\}\ , \tag{17}$$

with

$$a = \underline{r}_\lambda^\dagger \Gamma^{-1} \underline{r}_\lambda; \qquad b = \underline{s}^\dagger \Gamma^{-1} \underline{r}_\lambda; \qquad c = \underline{s}^\dagger \Gamma^{-1} \underline{s} \tag{18}$$

and $\underline{s}$ the vector formed by the filtered versions of the noise-free observation.

In the "noise only" case where $s(t) = 0$, some simplifications occur, leading to

$$f(\underline{r}) = \frac{1}{\pi^N |\det \Gamma|} \left| \prod_{n=1}^{N-1} \lambda_n \right|^2 \iint_{-\infty}^{+\infty} |y_1|^{2(N-1)} \exp\left(-a|y_1|^2\right)\, d\mathrm{Re}\{y_1\}\, d\mathrm{Im}\{y_1\} \tag{19}$$

and, finally, to

$$f(\underline{r}) = \frac{1}{\pi^N |\det \Gamma|} \left| \prod_{n=1}^{N-1} \lambda_n \right|^2 \frac{(N-1)!}{a^N}. \tag{20}$$

This result is quite general and can be applied to the specific situation of time-frequency reassignment by setting

$$h_1(t) = h^*(t)e^{i2\pi\nu t}; \qquad h_2(t) = (\mathcal{T}h^*)(t)e^{i2\pi\nu t}; \qquad h_3(t) = (\mathcal{D}h^*)(t)e^{i2\pi\nu t}, \tag{21}$$

thus leading to

$$\Gamma = 2\sigma^2 \begin{bmatrix} 1 & 0 & 0 \\ 0 & \Delta t_h^2 & -1/2 \\ 0 & -1/2 & 4\pi^2 \Delta \nu_h^2 \end{bmatrix}. \tag{22}$$

It has to be observed that $\det \Gamma = 2\sigma^6(16\pi^2\Delta t_h^2\Delta\nu_h^2 - 1)$, a quantity which, by virtue of Heisenberg's inequality, is always non-negative and is zero if and only if the window is Gaussian. In this case, Γ is no longer invertible and it is necessary to go back to the results of the previous Section.

Assuming that $h(t)$ is not Gaussian, we have to consider the reassignment vector

$$r = -\frac{1}{\Delta t_h}\text{Re}\left\{\frac{y_2}{y_1}\right\} + i\frac{1}{2\pi\Delta\nu_h}\text{Im}\left\{\frac{y_3}{y_1}\right\},\tag{23}$$

thus motivating the introduction of

$$\underline{r} = [r_1\ r_2]^t = \left[\frac{1}{\lambda_1}\frac{y_2}{y_1}\ \frac{1}{\lambda_2}\frac{y_3}{y_1}\right]^t,\tag{24}$$

with $\lambda_1 = -\Delta t_h$ and $\lambda_2 = 2\pi\Delta\nu_h$. The joint density of this vector can be shown to take on the form:

$$f(\underline{r}) = \frac{2}{\pi^2}\frac{H^2\left(H^2-1\right)^2}{\left[H^2\left(1+|r_1|^2+|r_2|^2\right)-2H\text{Re}\{r_1r_2^*\}-1\right]^3},\tag{25}$$

with $H = 4\pi\Delta t_h\Delta\nu_h$, whence the final result

$$f(r) = \iint_{-\infty}^{+\infty} f(\underline{r})\,d\text{Im}\{r_1\}\,d\text{Re}\{r_2\} = \frac{1}{\pi(1+|r|^2)^2},\tag{26}$$

which appears to be identical to the one obtained in the Gaussian case (see eq.(10)). We can remark that the frequency marginal of eq.(26) can be considered as well as the instantaneous frequency pdf of a colored bandpass Gaussian noise, with the consequence that its expression is in agreement with the result obtained in [5] by a different method.

The situation of "signal + noise" is far more complicated in the general case than in the Gaussian one and it will not be followed up here (see [2] for elements).

5. Conclusion

Statistical properties of reassignment vectors have been derived in simple, yet realistic and important situations. Given these results, it is now possible to incorporate statistical elements in the reassignment process, to validate its use in noisy situations and to increase the significance of the results it gives. It is also believed that this knowledge should help in extracting useful information directly from the reassignment vector field, in a statistical signal processing perspective. Finally, it is worthwhile to mention that the obtained results should be of help for providing a statistical basis to related techniques such as the "ridge and skeleton" method [6], whose principle has much to share with reassignment.

References

1. F. AUGER and P. FLANDRIN, "Improving the readability of time-frequency and time-scale representations by the reassignment method," *IEEE Trans. on Signal Proc.*, Vol. **SP-43**, No. 5, pp. 1068–1089, 1995.
2. E. CHASSANDE-MOTTIN, F. AUGER and P. FLANDRIN, "Statistique des vecteurs de réallocation du spectrogramme," Rapport Interne 96-01, Labo. de Physique, *ENS Lyon*, juillet 1996. A .ps file of this report can be obtained at the following address: http://www.physique.ens-lyon.fr/ts/report/r9601.ps.gz
3. B. PICINBONO, *Random Signals and Systems*, Prentice Hall, Englewood Cliffs, 1993.
4. A. PAPOULIS, *Probability, Random Variables and Stochastic Processes*, McGraw-Hill, New York, 1984.

5. H. BROMAN, "The instantaneous frequency of a Gaussian signal: the one-dimensional density function," *IEEE Trans. on Acoust., Speech and Signal Proc.*, Vol. **ASSP-29**, No. 1, pp. 103–111, 1981.
6. P. GUILLEMAIN and R. KRONLAND-MARTINET, "Characterization of acoustic signals through continuous linear time-frequency representations," *Proc. IEEE*, **84** (4), pp. 561–585, 1996.

Multidimensional Systems and Signal Processing, 9, 363–372 (1998)
© 1998 Kluwer Academic Publishers, Boston. Manufactured in The Netherlands.

Generalization of the Gram-Charlier/Edgeworth Series and Application to Time-Frequency Analysis

LEON COHEN*   leon.cohen@hunter.cuny.edu
Hunter College and Graduate Center of CUNY, 695 Park Ave., New York, NY 10021 USA

Received April 10, 1997; Accepted January 6, 1998

Abstract. We derive a general equation relating probability densities and as special cases we the obtain Gram-Charlier and Edgeworth series. This allows us to generalize these methods and clarify a number of issues pertaining to both probability theory and time-frequency analysis. In particular we show how the Gram-Charlier and Edgeworth series are related to the kernel method of time-frequency analysis. The approach allows us to construct densities that satisfy given constraints such as joint moments or conditional moments. Also, we show that the kernel has to be signal dependent and that to obtain a proper distribution it should be the ratio of two characteristic functions.

Key Words: Gram-Charlier Series, Edgeworth Series, probability theory, time-frequency analysis, characteristic function

1. Introduction and Main Results

The main purpose of the Gram-Charlier and Edgeworth series is to construct probability densities from partial information, such as a finite number of moments [7]. Our aim is to generalize these methods and to show the relationship with the kernel method of time-frequency analysis. This will allow us to present a new method for constructing time-frequency densities.

To express the results in a simple form we define the differential operator, D_x, and its adjoint, $D_x^{\dagger}$, by[1]

$$D_x = \frac{1}{j}\frac{d}{dx} \qquad D_x^{\dagger} = -\frac{1}{j}\frac{d}{dx} \tag{1.1}$$

Our main result is the following. Any two one dimensional densities, $P_1(x)$ and $P_2(x)$ are related by

$$P_2(x) = \phi(D_x^{\dagger})P_1(x) \tag{1.2}$$

where $\phi(\theta)$ is a function and where for the variable, θ, one substitutes the operator $D_x^{\dagger}$. We call $\phi(\theta)$ the kernel in analogy with its usage in time frequency. Furthermore, to obtain a proper density the kernel must be a functional of both distributions. In particular, the kernel is the ratio of the characteristic functions of the distributions

$$\phi(\theta) = \frac{M_2(\theta)}{M_1(\theta)} \tag{1.3}$$

---

* Work supported by the Office of Naval Research under contract No. N00014-97-1-0058, the NASA JOVE, and NSA HBCU/MI programs.

31

To prove Eq. (1.2) we first obtain a general operator relation involving the convolution theorem in Fourier analysis. We define Fourier transform pairs between $F(\theta)$ and $f(x)$ by

$$F(\theta) = \int f(x) e^{j x \theta} dx \qquad f(x) = \frac{1}{2\pi} \int F(\theta) e^{-j x \theta} d\theta \tag{1.4}$$

Now, suppose we have two functions, $f(x)$, $g(x)$ whose Fourier transforms are $F(\theta)$, $G(\theta)$ respectively. We ask for the function, $h(x)$, whose Fourier transform is the product $F(\theta)G(\theta)$. The answer is

$$h(x) = \frac{1}{2\pi} \int G(\theta) F(\theta) e^{-j x \theta} d\theta \tag{1.5}$$

$$= G(D^\dagger) f(x) \tag{1.6}$$

This relation is derived in Appendix 1.

We now prove Eq. (1.2). The characteristic function and probability density are Fourier transform pairs

$$M(\theta) = \int e^{j\theta x} P(x) dx \quad ; \quad P(x) = \frac{1}{2\pi} \int e^{-j\theta x} M(\theta) d\theta \tag{1.7}$$

Suppose we have two characteristic functions $M_1(x)$, $M_2(x)$ corresponding to two densities, $P_1(x)$ and $P_2(x)$. We write

$$M_2(\theta) = \frac{M_2(\theta)}{M_1(\theta)} M_1(\theta) = \phi(\theta) M_1(\theta) \tag{1.8}$$

Using Eq. (1.6) we immediately have Eq. (1.2).

Two Dimensions. Generalization to two and higher dimensions is straightforward. For the two dimensional case the characteristic function, $M(\theta, \tau)$, and density, $P(x, y)$, are related by

$$M(\theta, \tau) = \iint e^{j\theta x + j\tau y} P(x, y) \, dx \, dy \tag{1.9}$$

$$P(x, y) = \frac{1}{4\pi^2} \iint M(\theta, \tau) e^{-j\theta x - j\tau y} \, d\theta \, d\tau \tag{1.10}$$

As for the one dimensional case, if we have two densities, $P_1(x, y)$ and $P_2(x, y)$ with corresponding characteristic functions $M_1(\theta, \tau)$ and $M_2(\theta, \tau)$, we set

$$M_2(\theta, \tau) = \phi(\theta, \tau) M_1(\theta, \tau) \qquad \phi(\theta, \tau) = \frac{M_2(\theta, \tau)}{M_1(\theta, \tau)} \tag{1.11}$$

The same proof used for the one dimensional case leads to

$$P_2(x, y) = \phi(D_x^\dagger, D_y^\dagger) P_1(x, y) \tag{1.12}$$

2. Gram-Charlier and Edgeworth Series

Using Eq. (1.2) we derive the Gram-Charlier and Edgeworth results.

Edgeworth. If the characteristic function is written in terms of the cumulants κ_n's

$$M(\theta) = \exp\left[\sum_{n=1}^{\infty} \kappa_n \frac{j^n}{n!} \theta^n\right] \tag{2.1}$$

then

$$\phi(\theta) = \frac{M_2(\theta)}{M_1(\theta)} = \exp\left[\sum_{n=1}^{\infty} (\kappa_n^{(2)} - \kappa_n^{(1)}) \frac{j^n}{n!} \theta^n\right] \tag{2.2}$$

Using Eq. (1.2) we have that

$$P_2(x) = \exp\left[\sum_{n=1}^{\infty} (\kappa_n^{(2)} - \kappa_n^{(1)}) \frac{j^n}{n!} D_x^{\dagger n}\right] P_1(x) \tag{2.3}$$

Further simplification is possible but we will not do so in this paper. To obtain the standard Edgeworth result we specialize to the case where P_1 is the Normal density[2]

$$N(m, \sigma^2) = \sqrt{\frac{\alpha^2}{\pi}} \exp\left[-\alpha^2 (x-m)^2\right] \qquad ; \qquad \alpha^2 = \frac{1}{2\sigma^2} \tag{2.4}$$

For the Gaussian density, $\kappa_1 = m, \kappa_2 = \sigma^2$, and all higher cumulants are equal to zero. Hence

$$P_2(x) = \exp\left[j(\kappa_1^{(2)} - m)D_x^{\dagger} - \tfrac{1}{2}(\kappa_2^{(2)} - \sigma^2)D_x^{\dagger 2} + \sum_{n=3}^{\infty} \kappa_n^{(2)} \frac{j^n}{n!} D_x^{\dagger n}\right] N(m, \sigma^2) \tag{2.5}$$

This is one form of the Edgeworth expansion [7].

Gram-Charlier.[3] We first give the standard derivation. One expands, not the density but $e^{-\alpha^2(x-m)^2/2}$ times the density,

$$e^{-\alpha^2(x-m)^2/2} P(x) = \sum_{n=0}^{\infty} c_n u_n(x; m, \alpha) \tag{2.6}$$

where $u_n(x; m, \alpha)$ are the Hermit functions with parameters m, α, as defined by Eq. (8.4). Using the orthonormal properties one immediately obtains that the coefficients are given by

$$c_n = \sqrt{\frac{\alpha}{2^n n! \sqrt{\pi}}} \langle H_n(\alpha(x-m)) \rangle \tag{2.7}$$

and hence

$$P(x) = \frac{\alpha}{\sqrt{\pi}} e^{-\alpha^2(x-m)^2} \sum_{n=0}^{\infty} \frac{\langle H_n(\alpha(x-m)) \rangle}{2^n n!} H_n(\alpha(x-m)) \tag{2.8}$$

The usual Gram-Charlier expansion is presented for the standardized Gaussian, $m = 0, \alpha^2 = 1/2$, in which case

$$P(x) = \frac{\alpha}{\sqrt{\pi}} e^{-x^2/2} \sum_{n=0}^{\infty} \frac{\langle H_n(x/\sqrt{2}) \rangle}{2^n n!} H_n(x/\sqrt{2}) \tag{2.9}$$

This is the Gram Charlier type A series [7].

We now derive it in a more general way using Eq. (1.2). The derivation is more involved but is revealing and shows the generality of the method presented. Expand the characteristic function in terms of some functions, orthonormal or not,

$$M_1(\theta) = \sum_{n=0}^{\infty} a_n^{(1)} u_n^{(1)}(\theta) \quad ; \quad M_2(\theta) = \sum_{n=0}^{\infty} a_n^{(2)} u_n^{(2)}(\theta) \tag{2.10}$$

and set

$$\phi(\theta) = \frac{M_2(\theta)}{M_1(\theta)} = \sum_{n=0}^{\infty} q_n v_n(\theta) \tag{2.11}$$

Using Eq. (1.2) we have

$$P_2(x) = \sum_{n=0}^{\infty} q_n v_n(D^\dagger) P_1(x) \tag{2.12}$$

Now, suppose we take $v_n(\theta) = \theta^n$, and $P_1 = N(m, \sigma^2)$ to be the normal distribution, Eq. (2.4). Using Eq. (8.6) we then have

$$P_2(x) = \sum_{n=0}^{\infty} q_n D^{\dagger n} N(m, \sigma^2) = N(m, \sigma^2) \sum_{n=0}^{\infty} (-j\alpha)^n q_n H_n(\alpha(x-m)) \tag{2.13}$$

It remains to evaluate the q_n's. The characteristic function of the Gaussian is

$$M_1(\theta) = e^{jm\theta - \sigma^2\theta^2/2} = e^{jm\theta - \theta^2/(4\alpha^2)} \tag{2.14}$$

and one can straightforwardly show that

$$M_1(\theta) = \sum_{k=0}^{\infty} \frac{H_n(j\alpha)}{(2\alpha)^n n!} \theta^n \quad ; \quad \frac{1}{M_1(\theta)} = \sum_{k=0}^{\infty} \frac{H_n(m\alpha)}{(2\alpha j)^n n!} \theta^n \tag{2.15}$$

We write M_2 in terms of the moments,

$$M_2(\theta) = \sum_{n=0}^{\infty} \frac{j^n \langle x^n \rangle}{n!} \theta^n \tag{2.16}$$

to obtain

$$\phi(\theta) = \frac{M_2(\theta)}{M_1 \theta} = \sum_{n=0}^{\infty} \sum_{k=0}^{\infty} \frac{H_n(m\alpha)}{(2\alpha j)^n n!} \frac{j^k \langle x^k \rangle}{k!} \theta^{n+k} \tag{2.17}$$

$$= \sum_{n=0}^{\infty} \sum_{k=0}^{n} \frac{H_{n-k}(m\alpha)}{(2\alpha j)^{n-k}(n-k)!} \frac{j^k \langle x^k \rangle}{k!} \theta^n \tag{2.18}$$

Comparing with Eq. (2.11) we have

$$q_n = \sum_{k=0}^{n} \frac{H_{n-k}(m\alpha)}{(2\alpha j)^{n-k}(n-k)!} \frac{j^k \langle x^k \rangle}{k!} \tag{2.19}$$

$$= \frac{1}{(2\alpha j)^n n!} \sum_{k=0}^{n} \binom{n}{k} (-2\alpha)^k H_{n-k}(m\alpha) \langle x^k \rangle \tag{2.20}$$

or[4]

$$q_n = \frac{H_n(\alpha(x-m))}{(-j\alpha)^n 2^n n!} \tag{2.21}$$

Substituting in Eq. (2.13) we have

$$P_2(x) = N(m, \sigma^2) \sum_{n=0}^{\infty} \frac{\langle H_n(\alpha(x-m)) \rangle}{2^n n!} H_n(\alpha(x-m)) \tag{2.22}$$

If we specialize to the standardized Gaussian ($m = 0$, $\sigma^2 = 1$) we have

$$P_2(x) = N(0, 1) \sum_{n=0}^{\infty} \frac{\langle H_n(x/\sqrt{2}) \rangle}{2^n n!} H_n(x/\sqrt{2}) \tag{2.23}$$

Which is the same as Eq. (2.9).

3. Approximation

The basic aim of these series is that they allow one to construct densities satisfying partial information. To understand this we consider a simple example. Suppose we are given $\langle x \rangle$ and $\langle x^2 \rangle$ and wish to construct joint densities satisfying these values. We illustrate with both the Gram-Charlier and Edgeworth. For both cases we take P_1 to be the Normal distribution.

Gram-Charlier. If we keep only the first three terms in Eq. (2.22) and explicitly substitute for the Hermit polynomials we obtain

$$P_2(x) \sim N(m, \sigma^2)$$ (3.1)

$$\times \left[1 + \alpha^2(\langle x \rangle - m)(x - m) + \alpha^4(\langle x^2 \rangle - 2 \langle x \rangle m + m^2 - \tfrac{1}{2})(2(x-m)^2 - 1) \right]$$

Note that m, α are totally arbitrary.

Edgeworth. Keeping only the first three terms in the exponent of the Edgeworth series, Eq. (2.5), we have

$$P_2(x) = \exp\left[j(\kappa_1^{(2)} - m)D_x^\dagger - \tfrac{1}{2}(\kappa_2^{(2)} - \sigma^2)D_x^{\dagger 2} \right] N(m, \sigma^2)$$ (3.2)

Evaluation yields that

$$P_2(x) = N(\langle x \rangle, \eta^2) \qquad \eta^2 = \langle x^2 \rangle - \langle x \rangle^2$$ (3.3)

Hence for the Edgeworth series we get back a Normal density. The moments of the original Gaussian, m, σ^2 do not appear. However they would start to appear if we took higher terms in the expansion.

4. Time-Frequency Analysis

Historically there were a few time-frequency densities known, such as the Wigner and Rihaczek distribution. The author has given a general method to obtain an infinite number of them [1,2]. One takes

$$C(t, \omega) = \frac{1}{4\pi^2} \iiint s^*(u - \tfrac{1}{2}\tau) s(u + \tfrac{1}{2}\tau) \phi(\theta, \tau) e^{-j\theta t - j\tau\omega + j\theta u} \, du \, d\tau \, d\theta$$ (4.1)

where $\phi(\theta, \tau)$ is a two dimensional function called the kernel. Different densities are obtained by choosing different kernels. The importance of the method is that the kernel determines the properties of the distributions and hence one can study and selectively obtain densities with certain properties by imposing constraints on the kernel. Much of the work done with this method has been with the so called bilinear case which is obtained by taking kernels that are not functionals of the signal. In that case the signal enters twice and hence is called bilinear. However there has been considerable work with the non-bilinear case.

We believe the considerations in the previous section sheds light on the kernel method and also provides us a new means of obtaining distributions with properties that we want to impose. Let use rewrite Eq. (4.1) in the following form

$$C(t, \omega) = \frac{1}{4\pi^2} \iint M(\theta, \tau) e^{-j\theta t - j\tau\omega} \, d\theta \, d\tau$$ (4.2)

where

$$M(\theta, \tau) = \phi(\theta, \tau) A(\theta, \tau)$$ (4.3)

$$A(\theta, \tau) = \int s^*(u - \tfrac{1}{2}\tau) s(u + \tfrac{1}{2}\tau) e^{j\theta u} \, du$$ (4.4)

and where $A(\theta, \tau)$ is the symmetrical ambiguity function. Superficially Eq. (4.3) looks like Eq. (1.11). However, there are two significant differences. First, $A(\theta, \tau)$ is not a proper characteristic function except for a Gaussian signal. Secondly, as mentioned above the kernel has traditionally been taken to be a pure function and that generally cannot be the case. However, as we will discuss in another paper the above formulation can be viewed as an approximation.

Of particular interest has been the question of positivity. It has always been a difficult issue in time-frequency analysis to understand what the conditions on the kernel are to achieve positivity. We now see why. The kernel can not in general be a pure function but a functional of the distributions involved.

5. An Example

To illustrate how the above methods can be used in time-frequency analysis we illustrate with an example involving conditional quantities. Suppose we are given conditional moments, how can we construct a joint density having these moments? One way to approach the problem is to write the joint as

$$P(t, \omega) = P(t)P(\omega|t) \tag{5.1}$$

where $P(t)$ is the time marginal and $P(\omega|t)$ is the conditional density of frequency for a given time. We have previously argued that if the signal is written in terms of its amplitude and phase

$$s(t) = A(t) e^{j\varphi(t)} \tag{5.2}$$

then plausible first and second conditional moments and conditional standard deviations for frequency at a given time are

$$\langle \omega \rangle_t = \varphi'(t) \tag{5.3}$$

$$\langle \omega^2 \rangle_t = \left(\frac{A'(t)}{A(t)} \right)^2 + \varphi'^2(t) \tag{5.4}$$

$$\sigma^2_{\omega|t} = \left(\frac{A'(t)}{A(t)} \right)^2 \tag{5.5}$$

If we use the Edgeworth series, Eq. (3.3), and take the time marginal to be $|s(t)|^2$ we obtain

$$P(t, \omega) \sim A^2(t)\sqrt{\frac{1}{2\pi \sigma^2_{\omega|t}}} \exp\left[-\frac{(\omega - \varphi'(t))^2}{2\sigma^2_{\omega|t}} \right] \tag{5.6}$$

If we use the Gram-Charlier, Eq. (3.1), we have that

$$P(t, \omega) \sim A^2(t)N(m, \sigma^2) \left[1 + \alpha^2(\varphi'(t) - m)(\omega - m) \right. \tag{5.7}$$

$$\left. + \alpha^4(\sigma^2_{\omega|t}\varphi'^2(t) - 2\varphi'(t)m + m^2 - \tfrac{1}{2})(2(\omega - m)^2 - 1) \right] \tag{5.8}$$

6. Conclusion

We believe we have clarified a number of issue both pertaining to statistics and time-frequency densities. We have shown that the Gram-Charlier and Edgeworth Series are special cases a general operator relation between distributions as given by Eq. (1.2). In regard to time-frequency analysis our results show is that the kernel must be a functional of the signal and hence the resulting distributions must be non-bilinear. If the kernel is the ratio of two characteristic functions then we are assured positivity. We point out that there is a general method to generate positive distributions for a given set of marginals but the formulation is quite different then that given here [3,4,2]. Recently a major advance in this regard was made by Loughlin et. all. where they have obtained maximum entropy densities that satisfy the marginals and have derived a number of interesting results based on requiring positivity of the density [5,6]. Relation of these methods with the formulation developed here will be discussed in a subsequent paper.

7. Appendix 1

We now give two proofs of Eq. (1.6). In Eq. (1.5), expand $G(\theta)$ in a Taylor series

$$G(\theta) = \sum_{n=0}^{\infty} g_n \theta^n \tag{7.1}$$

to obtain

$$h(x) = \frac{1}{2\pi} \sum_{n=0}^{\infty} g_n \int \theta^n F(\theta) e^{-jx\theta} \, d\theta \tag{7.2}$$

$$= \frac{1}{2\pi} \sum_{n=0}^{\infty} g_n \int D_x^n F(\theta) e^{-jx\theta} \, d\theta \tag{7.3}$$

$$= \frac{1}{2\pi} \sum_{n=0}^{\infty} g_n D_x^n \int F(\theta) e^{-jx\theta} \, d\theta \tag{7.4}$$

$$= G(D^{\dagger}) f(x) = G(-D) f(x) \tag{7.5}$$

Alternatively we can derive this result from the convolution theorem. The function, $h(x)$, whose Fourier transform is the product $F(\theta)G(\theta)$ is

$$h(x) = \frac{1}{2\pi} \int G(\theta) F(\theta) e^{-jx\theta} \, d\theta \tag{7.6}$$

$$= \int g(x') f(x - x') \, dx' \tag{7.7}$$

But

$$f(x - x') = e^{-x'\frac{d}{dx}} f(x) = e^{jx'D_x^{\dagger}} f(x) \tag{7.8}$$

and therefore

$$h(x) = \int g(x') f(x - x') \, dx' \tag{7.9}$$

$$= \int g(x') e^{jx' D_x^{\dagger}} f(x) \, dx' \tag{7.10}$$

$$= \int g(x') e^{jx' D_x^{\dagger}} \, dx' f(x) \tag{7.11}$$

$$= G(D_x^{\dagger}) f(x) \tag{7.12}$$

8.   Appendix 2

We use the standard definition for the Hermit functions

$$u_n(x) = \frac{1}{\sqrt{2^n n! \pi^{1/4}}} H_n(x) e^{-x^2/2} \tag{8.1}$$

where H_n are the Hermit polynomials[5]

$$H_n(x) = \sum_{k=0}^{[n/2]} \frac{(-1)^k n!}{k!(n-2k)!} (2x)^{n-2k} = (-1)^n e^{x^2} \frac{d^n}{dx^n} e^{-x^2} \tag{8.2}$$

The Hermite functions are orthonormal

$$\int u_n(x) u_m(x) \, dx = \delta_{nm} \tag{8.3}$$

Also the functions

$$u_n(x; m, \alpha) = \sqrt{\frac{\alpha}{2^n n! \sqrt{\pi}}} H_n(\alpha(x - m)) e^{-\alpha^2 (x-m)^2/2} \tag{8.4}$$

form an orthonormal complete set and

$$H_n(\alpha(x - m)) = \frac{(-1)^n}{\alpha^n} e^{\alpha^2 (x-m)^2} \frac{d^n}{dx^n} e^{-\alpha^2 (x-m)^2} \tag{8.5}$$

$$D^{\dagger n} N(m, \sigma^2) = (-j\alpha)^n H_n(\alpha(x - m)) N(m, \sigma^2) \tag{8.6}$$

Notes

1. In probability theory the operator D_x is usually defined by $D_x = \frac{d}{dx}$. We prefer our definition because it makes it a Hermitian operator.
2. If we had started by immediately taking P_1 to be the Normal distribution the derivation given here would be almost identical to the one given in the probability literature [7]. Eq. (2.3) is more general since it involves any two densities.

3. We will need some properties of the Hermite polynomials. These are given in appendix 2.

4. We have used the fact that $H_n(-x) = (-1)^n H_n(x)$ and that $H_n(x+y) = \sum_{k=0}^{n} \binom{n}{k} H_{n-k}(x)(2y)^k$.

5. This definition of the Hermite polynomials is the standard one. However in the probability literature the Hermit polynomials are defined slightly differently: $\mathcal{H}_n(x) = (-1)^n e^{x^2/2} \frac{d^n}{dx^n} e^{-x^2/2}$. The relation between the two is $H_n(x) = 2^{n/2} \mathcal{H}_n(\sqrt{2}x)$

References

1. L. Cohen, "Generalized Phase-Space Distribution Functions," *Jour. Math. Phys.*, vol. 7, 1966, pp. 781–786.

2. L. Cohen, *Time-Frequency Analysis*, Prentice-Hall, 1995.

3. L. Cohen and Y. Zaparovanny, "Positive Quantum Joint Distributions," *J. Math. Phys.*, vol. 21, 1980, pp. 794–796.

4. L. Cohen and T. Posch, "Positive Time-Frequency Distribution Functions," *IEEE Trans. ASSP*, vol. 33, 1985, pp. 31–38.

5. P. Loughlin, J. Pitton and L. E. Atlas, "Construction of Positive Time-Frequency Distributions," *IEEE Trans. Sig. Proc.*, vol. 42, 1994, pp. 2697–2705.

6. P. Loughlin and B. Tacer, "On the Amplitude- and Frequency-Modulation Decomposition of Signals," *J. Acoust. Soc. Amer.*, vol. 100, no. 3, 1996, pp. 1594–1601.

7. A. Stuart and K. K. Ord, *Kendall's Advanced Theory of Statistics*, Halsted Press, 1994.

Multidimensional Systems and Signal Processing, 9, 373–381 (1998)
© 1998 Kluwer Academic Publishers, Boston. Manufactured in The Netherlands.

Time-Frequency Analysis Using Local Taylor Series Expansion

BENJAMIN FRIEDLANDER*
Department of Electrical and Computer Engineering, University of California, Davis, CA 95616, USA

ERIK FRIEDLANDER
Signal Processing Technology, Ltd., Palo Alto, CA 94306, USA

Received August 16, 1996; Revised July 9, 1997

Abstract. We describe an approach to time-frequency analysis based on the local approximation of the signal by a first order Taylor series. We show that the Taylor approximation provides a representation of the signal in terms of its instantaneous frequency and instantaneous bandwidth. This representation can be translated into the frequency domain in a straightforward manner. The key to this approach is the *local* decomposition of the signal into its components, which is similar to the problem of estimating the parameters of complex exponentials from observation of their sum. The resulting time-frequency representation (TFR) does *not* have the time and frequency marginal properties shared by many of the time-frequency distributions presented in the literature, but is additive over the signal components and, by its construction, does not have cross-terms.

Key Words: time, frequency, spectrum, signal analysis

I. Introduction

Most of the work on time-frequency representation (TFR) is based on the idea of defining the TFR as a joint time-frequency distribution $P(t, w)$, which provides a measure of the energy density of the signal at a particular time and frequency [1]–[6]. These distributions are assumed to have the time and frequency marginal properties:

$$P_t(t) = \int_{-\infty}^{\infty} P(t, w)dw = |s(t)|^2 \qquad (1)$$

and

$$P_w(w) = \int_{-\infty}^{\infty} P(t, w)dt = |S(w)|^2 \qquad (2)$$

There are, of course, infinitely many distributions $P(t, w)$ which can be constructed for a given set of marginals. Much of the current work on TFR involves attempts to select a particular distribution (or class of distributions) which will be consistent with the marginals, and in addition, will have other desirable properties (positivity, time and frequency shift, scale invariance, etc.) or a particular structure (e.g. a bilinear function of the signal). The constraints imposed by requiring a sufficient number of other properties or structures, make it possible to uniquely determine a time-frequency distribution.

---

* This work was supported by the Office of Naval Research under contract No. N00014-95-1-0912.

41

One of the consequences of requiring the marginal properties to hold, is that the TFR is not additive with respect to the components of the signal. In other words, if a signal $s(t)$ is composed of M components $s(t) = \sum_{i=1}^{M} s_i(t)$, then in general $P(t, w) \neq \sum_{i=1}^{M} P_i(t, w)$, where $P_i(t, w)$ is the TFR of $s_i(t)$, as discussed, e.g., in [6], sec. 1.10. This lack of additivity usually manifests itself in undesirable "cross-terms," which have been the subject of considerable discussion.

In this paper we explore an approach to TFR which has, by its definition, the additivity property mentioned above. We believe that preserving additivity is useful, perhaps even essential, for time-frequency analysis of multi-components signals. Consider, in particular, a signal composed of components from several physically unrelated phenomena. Any reasonable interpretation of the TFR as a physically meaningful time-frequency energy distribution, would require that the TFR of the composite signal be the sum of the TFRs of the (physically unrelated) components.

The proposed approach is based on the local approximation of the signal by a first order Taylor series. In the case of a single component signal, this approximation is straightforward, as shown in section 2. The coefficients of the Taylor expansion are the instantaneous frequency and bandwidth, from which a TFR can be easily constructed.

The analysis of a multi-component signal proceeds by first decomposing it (locally) into its components, and then treating each of the components as in the case of a single-component signal. This procedure is described in section 3.

II.   Local Expansion for Mono-Component Signals

Frequency analysis has to do with discovering periodicities in signals which have repeatable patterns. Time-frequency analysis attempts to extend this notion to nonstationary signals which do not necessarily have any periodicities or repeatable patterns. To do this we will consider time-frequency analysis as an attempt to extend the local properties of the signal to the entire observation interval.

Consider for example a mono-component analytic signal of the form

$$s(t) = a(t)e^{j\phi(t)} \tag{3}$$

observed over the interval $[-T/2, T/2]$. We are observing the signal at time t and want to get a simple approximation of the signal in that neighborhood. We will use a first order Taylor expansion

$$s(t + \tau) \approx s(t) + \dot{s}(t)\tau \tag{4}$$

where

$$\dot{s}(t) = \left(\frac{\dot{a}(t)}{a(t)} + j\dot{\phi}(t)\right) a(t)e^{j\phi(t)} \tag{5}$$

or

$$\dot{s}(t) = \left(\frac{\dot{a}(t)}{a(t)} + j\dot{\phi}(t)\right) s(t) \tag{6}$$

Thus, locally we can represent the signal as

$$s(t + \tau) = \left(1 + \frac{\dot{a}(t)}{a(t)}\tau + j\dot{\phi}(t)\tau\right)s(t) \tag{7}$$

or, if we assume that $|\frac{\dot{a}(t)}{a(t)}\tau| \ll 1$ and $|\dot{\phi}(t)\tau| \ll 1$, and use the approximation $e^x \approx 1 + x$,

$$s(t + \tau) \approx s(t)e^{j\dot{\phi}(t)\tau + \frac{\dot{a}(t)}{a(t)}\tau} \tag{8}$$

Let us define

$$\beta(t) = \beta_R(t) + j\beta_I(t) = \frac{\dot{a}(t)}{a(t)} + j\dot{\phi}(t) \tag{9}$$

where the imaginary part of $\beta(t)$ is proportional to the instantaneous frequency $\dot{\phi}(t)$, and its real part is the instantaneous bandwidth $|\frac{\dot{a}(t)}{a(t)}|$ (cf. equation (13.2) in [6]). Then,

$$s(t + \tau) \approx s(t)e^{\beta(t)\tau} \tag{10}$$

Equation (10) captures the local behavior of the analytic signal $s(t)$, in terms of its instantaneous frequency and instantaneous bandwidth.

Next let us extend this approximation from the neighborhood of t to a larger interval. For example, we define

$$s(t, \tau) = s(t)e^{\beta(t)\tau}, \quad 0 \le \tau \le T \tag{11}$$

Other choices of this interval are possible [7].

Since this extension does not preserve the local energy of the signal, we will modify the definition above by adding an appropriate normalization, as follows. Let

$$s(t, \tau) = s(t)e^{\beta(t)\tau}/\sqrt{e(t)}, \quad -T/2 \le \tau \le T/2 \tag{12}$$

$$e(t) = \int_{-T/2}^{T/2} |e^{\beta(t)\tau}|^2 \, dt \tag{13}$$

or

$$e(t) = \frac{\exp\{T\beta_R(t)\} - \exp\{-T\beta_R(t)\}}{2\beta_R(t)} = T\frac{\sinh\{T\beta_R(t)\})}{T\beta_R(t)} \tag{14}$$

With this normalization we have

$$\int_{-T/2}^{T/2} |s(t, \tau)|^2 \, dt = |s(t)|^2 \tag{15}$$

Finally, we define the time frequency representation $P(t, w)$ as the magnitude squared of the Fourier transform of $s(t, \tau)$ with respect to τ

$$P(t, w) = \left| \int_{-T/2}^{T/2} h(\tau)s(t, \tau)e^{-jw\tau} \, d\tau \right|^2 \tag{16}$$

where $h(\tau)$ is an optional windowing function. Equivalently,

$$P(t, w) = (|s(t)|^2/e(t)) \left| \int_{-T/2}^{T/2} h(\tau)e^{(\beta(t)-jw)\tau} \, d\tau \right|^2 \tag{17}$$

In the case of a rectangular window we get

$$P(t, w) = (|s(t)|^2/e(t))(T^2/4) \left| \frac{\sinh\{(\beta(t) - jw)T/2\}}{(\beta(t) - jw)T/2} \right|^2 \tag{18}$$

In the case of a constant amplitude signal $\beta_r(t) = 0$ and therefore $e(t) = 1$ and (18) simplifies to

$$P(t, w) = |s(t)|^2(T^2/4) \left| \frac{\sin\{(\beta_I(t) - w)T/2\}}{(\beta_I(t) - w)T/2} \right|^2 \tag{19}$$

In other words, the TFR will consist of a narrow line centered on the trajectory of the instantaneous frequency. The width of the line is on the order of $1/T$, and the sidelobe structure is that of the Dirichlet kernel. Using other windows will modify the sidelobes and mainlobe in the usual manner.

A. Discussion

1. The TFR defined in (18) has by its construction the time-marginal property. Therefore, the integral over the entire time-frequency plane will yield the total signal energy $E = \int_{-T/2}^{T/2} |s(t)|^2 \, dt$

2. In general the TFR will not have the frequency marginal property.

3. Note that the signal can be reconstructed from $\beta(t)$ as long as we assume that the amplitude function $a(t)$ is positive, since it follows from equation (6) that

$$\frac{\dot{s}(t)}{s(t)} = \beta(t) \tag{20}$$

which can be integrated to yield

$$s(t) = s(t_0) \exp \left\{ \int_{t_0}^{t} \beta(\tau)d\tau \right\} \tag{21}$$

where t_0 is some initial time point.

In other words, for this class of signals, the instantaneous frequency $\beta_I(t)$ and the instantaneous bandwidth (with the sign preserved) $\beta_R(t)$ fully define the signal, up to an initial value. We may consider the instantaneous frequency and bandwidth to be an alternative parameterization of the signal, and the TFR, a straightforward translation of this parameterization to the frequency domain.

III. The Multi Component Case

Having defined the Taylor-based TFR for mono-component signals, we now proceed to extend it to the more challenging multi-component case. In other words, we assume that the signal consist of a sum of several components of the form considered in the previous section, i.e.

$$s(t) = \sum_{m=1}^{M} s_m(t) \tag{22}$$

where $a(t)$ and $\phi(t)$ are smooth functions. The number of components is not necessarily fixed over time. Components may start and end at different times. Therefore, we should allow for the possibility of a time varying number of components, which we denote by $M(t)$, in which case,

$$s(t) = \sum_{m=1}^{M(t)} a_m(t) e^{j\phi_m(t)} \tag{23}$$

Let us assume for a moment that we were able to decompose the signal in this fashion. Then we could define the TFR of the composite signal as the sum of the TFRs of the components,

$$P(t, w) = \sum_{m=1}^{M(t)} P_m(t, w) \tag{24}$$

where $P_m(t, w)$ is the TFR of the m-th component, using the definition introduced in the previous section, i.e.

$$P_m(t, w) = (|s_m(t)|^2 / e_m(t)) \left| \int_{-T/2}^{T/2} h(\tau) e^{(\beta_m(t) - jw)\tau} \, d\tau \right|^2 \tag{25}$$

where

$$\beta_m(t) = \frac{\dot{a}_m(t)}{a_m(t)} + j\dot{\phi}_m(t) \tag{26}$$

In other words, the signal $s(t)$ is characterized at time t by a set of $M(t)$ instantaneous frequencies and bandwidths $\{\beta_m(t)\}$. The TFR is just a translation of this set of parameters into a time-frequency domain picture.

Note that this TFR is additive in the components and its time-marginal is the sum of the time-marginals of the components. Thus, the composite TFR will *not* have a time-marginal property since

$$|s(t)|^2 \neq \sum_{m=1}^{M(t)} |s_m(t)|^2 \tag{27}$$

The key to this approach is the decomposition of a signal into a sum of analytic signals as indicated in equation (22). Attempts to perform the decomposition using only the signal and its derivatives at a single time point t fails to produce any useful results. Instead we will use values of $s(t)$ and its derivative $\dot{s}(t)$ over a small neighborhood of t.

Using the first order Taylor expansion in equation (10) we get

$$s(t + \tau) = \sum_{m=1}^{M(t)} s_m(t + \tau) \approx \sum_{m=1}^{M(t)} s_m(t)e^{\beta_m(t)\tau} \tag{28}$$

Let us assume that $\beta(t)$ varies sufficiently slowly so that the approximation above holds well over a small interval $[t, t + \Delta]$. The problem of performing the decomposition is equivalent to estimating $2M(t)$ unknowns $\{s_m(t), \beta_m(t)\}$ from the signal $s(t+\tau)$, $\tau \in [t, t+\Delta]$. Note that this problem is the same as estimating the parameters of a sum of complex exponentials,

$$s(t + \tau) = \sum_{m=1}^{M(t)} c_m d_m^\tau \tag{29}$$

where d_m are the exponentials, and c_m are their amplitudes. This problem has been studied extensively, starting with Prony [8], and various techniques are available to perform the estimation. In our experiments we used a modification of a method due to Kumaresan and Tufts, which is described in [9].

Having estimated the exponential parameters, the TFR can be easily computed as before,

$$P_m(t, w) = |c_m(t)|^2/e_m(t) \left| \int_{-T/2}^{T/2} h(\tau)d_m(t)^\tau e^{-jw\tau} \, d\tau \right|^2 \tag{30}$$

where $h(\tau)$ is an optional window,

$$e_m(t) = \int_{-T/2}^{T/2} |d_m(t)^\tau|^2 \, dt \tag{31}$$

and finally

$$P(t, w) = \sum_{m=1}^{M(t)} P_m(t, w) \tag{32}$$

A. Discussion

The formulation above emphasizes the fundamental role of the *local* decomposition of a signal into its components in defining and understanding the time frequency representation of nonstationary signals. This decomposition involves many issues which are not discussed here, due to space limitation, such as: (i) constraints on the components which are needed to ensure the uniqueness of the decomposition; (ii) how to determine the number of components; (iii) how to determine the size of the neighborhood over which the decomposition is performed; and (iv) characterization of the time-frequency resolution of the TFR. These and related issues are discussed in [7].

We note that the procedure outlined in this paper is related to the idea of performing time-frequency analysis by replacing the Fourier transform (in a short-time Fourier transform) by an autoregressive "high-resolution" spectral analysis technique. The framework presented here provides the "correct" way of using an autoregressive modeling technique.

B. Some Examples

Figures 1 and 2 depict the TFR produced by the procedures described above, for a mono-component and two-component signal.

The first example corresponds to a signal consisting of a sequence of decaying exponentials whose amplitudes vary linearly with time (i.e. the signal envelope is triangular). It is interesting to note that even for this single component signal this TFR is quite different

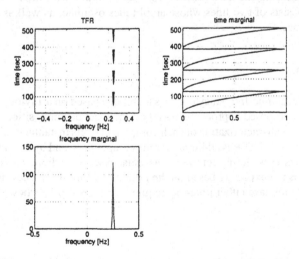

Figure 1. The TFR, time and frequency marginals for a complex exponential with a triangularly varying amplitude, using the "intuitive" TFR.

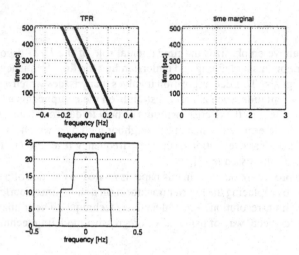

Figure 2. The TFR, time and frequency marginals for the sum of two constant unit amplitude chirps, using the intuitive TFR.

from TFRs based on time-frequency distributions which obey the marginal constraints, see e.g. figure 14.2 on page 203 in [6], for comparison.

The second example corresponds to a signal consisting of two unit amplitude linear FM chirps, with the same chirp rate. Note that the TFR consists of two constant amplitude lines depicting the instantaneous frequency trajectories of the two chirps. The "usual" TFR for this signal consists of two lines whose amplitudes oscillate, as well as a "cross-term" in-between the lines.

IV. Conclusions

We have introduced a time-frequency analysis technique based on a Taylor series expansion of the signal. The proposed approach is based on the local decomposition of a signal into components, and the characterization of each component by its instantaneous frequency and bandwidth. The resulting TFR is additive over the components, and by its construction, does not have the type of cross-terms commonly associated with quadratic and other nonlinear TFRs. This TFR is not explicitly based on the notion of a time-frequency distribution, and in particular, does not have either time- or frequency-marginal properties.

References

1. L. Cohen, "Time-Frequency Distributions—A Review," *Proc. IEEE*, vol. 77, no. 7, 1989, pp. 941–981.

2. F. Hlawatsch and G. F. Boudreaux-Bartels, "Linear and Quadratic Time-Frequency Signal Representation," *IEEE Signal Processing Magazine*, 1992, pp. 21–67.

3. B. Boashash, "Estimating and Interpreting the Instantaneous Frequency of a Signal—Part 1: Fundamentals," *Proc. IEEE*, vol. 80, no. 4, 1992, pp. 520–538. "Part II: Algorithms and Applications," pp. 539–568.

4. B. Boashash (ed.), *New Methods in Time Frequency Analysis*, Sydney, Australia: Longman Chesire, 1992.

5. *Proceedings of the 1992 Symposium on Time-Frequency and Time-Scale Analysis*, Victoria, British Columbia, Canada, October 4–6, 1992.

6. L. Cohen, *Time-Frequency Analysis*, Prentice-Hall, 1995.

7. B. Friedlander, "Time-Frequency Analysis for Multi-Component Signals," in preparation.

8. R. Prony, "Essai Experimental et Analytique, etc.," *L'ecole Polytechnique, Paris*, vol. 1, 1795, pp. 24–76.

9. B. Porat and B. Friedlander, "A Modification of the Kumaresan-Tufts Method for Estimating Rational Impulse Responses," *IEEE Trans. Acoustics, Speech and Signal Processing*, vol. ASSP–34, no. 5, 1986, pp. 1336–1338.

Multidimensional Systems and Signal Processing, 9, 383–389 (1998)

The Analysis of Under Water Vehicle Non-Stationary Acoustic Transient Signals Using a New Fast Algorithm for Constructing Positive Time-Frequency Distributions

DALE GROUTAGE
Naval Surface Warfare Center, Bremerton Detachment, Puget Sound, Bremerton, WA 98314-5215

Received December 30, 1996; Revised May 1, 1997

Abstract. It is well know that under water vehicles emit many types of acoustic signals, including stationary as well as non-stationary. The non-stationary transient signals are receiving considerable attention, especially from the standpoint of detection and tracking of under water vehicles. Thus, modern methods for processing these non-stationary acoustic transient signals are required. This paper presents a very powerful method for analyzing acoustic transient signals, the positive time-frequency distribution. A fast algorithm that implements the minimum cross-entropy positive time-frequency distribution makes practical the processing of "real world data", like under water vehicle acoustic transient signals. An example of such a signal is presented, which is a generic acoustic transient signal from a under water vehicle. The signal is only representative, as it is normalized in time, in frequency, and in amplitude. The positive time-frequency distribution constructed for this generic transient signal is contrasted with the one-third octave method, which is currently the primary method being used by the under water vehicle community to analyze under water vehicle transients. The positive distribution is also contrasted with broad band and narrow band spectrograms.

Key Words: non stationary signals, acoustic transients, time-frequency underwater vehicles, positive time-frequency distributors

I. Introduction

Background—Under water vehicle Transient Signals

Stationary, steady-state acoustic signatures for modern under water vehicles, which are characterized by specific features such as frequency lines, frequency bands, specific bandwidths, modulation, etc. have been reduced to have very low energy level content. This means that under water vehicles are quiet with regard to their steady-state signatures. However, non-stationary signatures for under water vehicles do not share this same level of "quiet character", and as a result are a primary element for detection. Thus, a high priority is being placed on quantifying the non-stationary signatures of under water vehicles because of the importance that stealth plays in the mission of the modern day under water vehicle.

The current method for transient analysis, short term ($\sim$10 ms) one-third octave analysis, does little more than provide a rough estimate of the energy involved. What is required is a signal processing technique that is capable of characterizing transient events so source identification can be accomplished. Time-frequency representations possess the greatest potential to obtain acoustic features that would "fingerprint" the transient event. The positive time-frequency distribution provides a meaningful representation of the joint time-frequency energy density. This is precisely what is required for under water vehicle acoustic

51

transient signal analysis, because bilinear time-frequency techniques that yield negative energy densities are meaningless to the understanding of the under water vehicle detection and detectability problem. Furthermore, short-time techniques, such as the spectrogram and wavelets, misrepresent the energy density of transient signals and therefore, are also limited in their usefulness to understanding under water vehicle detection and detectability issues.

Background—Technical

The development of modern techniques to process non-stationary signals has and continues to be the focus of much research where the goal is the description of the distribution of signal energy as a joint function of time and frequency. Ideally, such a distribution should be manifestly positive for a proper interpretation as a joint energy density function and, further, should yield the correct marginal densities of time $|s(t)|^2$ and frequency $|S(f)|^2$. Short-time methods (e.g. spectrograms, wavelets) can not achieve this goal, and many believed it was impossible. In 1966, Cohen [1] developed a unifying formulation for time-frequency distributions with correct marginals; however, the condition for positivity was not known until several years later, Cohen with Posch [2], and Cohen with Zaparovanny [3]. The Cohen formulation stipulates that *all* time-frequency representations can be obtained from

$$Q(t, f) = \iiint \left\{ s^*\left(u - \frac{1}{2}\tau\right) s\left(u + \frac{1}{2}\tau\right) \phi(\theta, \tau) e^{-j\theta t - j\tau 2\pi f + j\theta u} \right\} du\, d\tau\, d\theta$$

$$(1)$$

where $\phi(\theta, \tau)$ is a two-dimensional function, called the kernel, which determines the distribution and its properties. For a time-frequency distribution $Q(t, f)$ of the signal $s(t)$ (real or complex) to be interpreted as a joint energy density, it must, at the very least, satisfy the two fundamental properties of nonnegativity and the correct marginals for all time and frequency:

$$Q(t, f) \geq 0 \tag{2}$$

$$\int_{-\infty}^{\infty} Q(t, f)dt = |S(f)|^2 \tag{3a}$$

$$\int_{-\infty}^{\infty} Q(t, f)df = |s(t)|^2 \tag{3b}$$

where $S(f) = \int_{-\infty}^{\infty} s(t)e^{-j2\pi ft}dt$ is the Fourier transform of the signal.

The original minimum cross-entropy positive time-frequency algorithm of Loughlin, Pitton, and Atlas [4], which constructs the distribution $Q(t, f)$, places two constrains on the problem; (1) the amount of random access memory required to store the distribution $Q(t, f)$, and (2) the number of computations required to construct it. To address these two computational requirements imposed by the original algorithm, a fast algorithm was developed by Groutage [5].

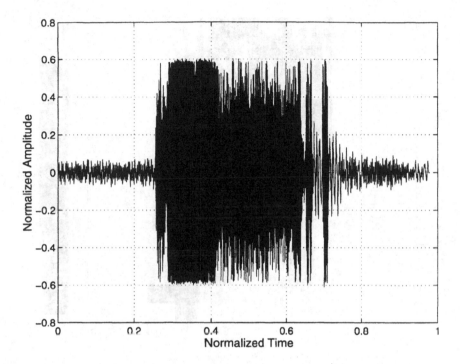

Figure 1. Time series.

II.   Example—Under Water Vehicle Non-Stationary Generic Transient Acoustic Signal

The following example presents the analysis of a generic, non-stationary acoustic transient signal of a under water vehicle. The fast minimum cross-entropy positive time-frequency distribution algorithm is used to process the time series illustrated in Figure No. 1.   This time series, which contains 32000 data points, represents the non-stationary acoustic transient signal. Note that this signal is generic in that both time and amplitude are normalized. The goal of the analysis is to define the exact nature of the signal such that scientists can relate features of the signal to physical systems that cause the signal to occur. Precisely in time when events occur and the frequency make up of the events are crucial to the diagnostic process. Thus, the analysis process must attribute simultaneously high resolution in both time and frequency. The images of Figures 2 through 5 are illustrated in dB levels. The bar scale on the right side of each image depicts the dB level per color band. The one-third octave image illustrated in Figure No. 2 represents the analysis process that is currently being utilized by the under water vehicle community to process non-stationary acoustic transient signals from under water vehicles. At best, the one-third octave analysis method provides a gross estimate of energy level over broad frequency bands. It is quite evident that resolution

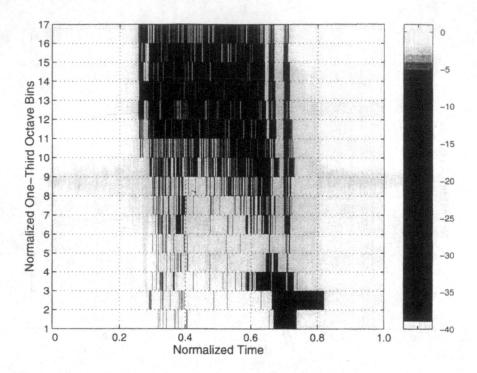

Figure 2. One-third octave representation.

in both time and frequency is very poor! In contrast, the broad band spectrogram (FFT size is 2048 with window length of 64) presented in the image of Figure No. 3 provides the high resolution in time required to identify when in time the events occur. However, note that resolution in frequency is very poor. Energy is smeared across the entire energy band up to the Nyquist frequency. On the other hand, the image of Figure No. 4, the narrow band spectrogram (FFT size of 2048 with window length of 1024), exhibits excellent resolution in frequency, but smears energy in time.

The problem with the spectrogram is that high resolution cannot be achieved in both time and frequency with a single processing window length (short window is broad band spectrogram, long window is narrow band spectrogram). The dilemma of short-time techniques is that depending on the window chosen, short-time techniques give very different answers: one suggests broadband transients, the other indicates narrowband tones. The positive time-frequency distribution, which is not a short-time technique, gives an accurate picture of the joint energy structure of the signal. The image of Figure No. 5, depicting the positive distribution, does contain a proper joint time-frequency energy distribution. It exhibits high resolution in both time and frequency, and of course, both time and frequency marginals are satisfied.

Note that only in the positive distribution of Figure 5 are the discrete frequency steps

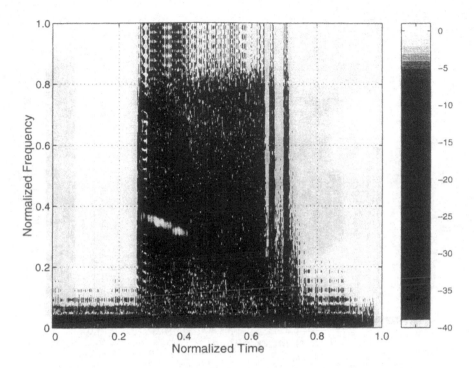

Figure 3. Broad band spectrogram.

uniquely discernible in the down-sweep frequency modulated feature of the signal (bright yellow feature in figures between normalized frequencies 0.2 and 0.4). This feature is completely misrepresented in the spectrograms of figures 3 and 4. It is interesting to note that in Figure 3, the broad band spectrogram, the down sweep frequency modulated feature appears as a very broad band of frequencies that are decreasing in frequency as time progresses. In Figure 4, the narrow band spectrogram, this same down sweep frequency modulation feature appears as if it is an "intertwined rope feature". This is a very unique artifact that is introduced by the spectrogram!

Under realistic computational requirements, a fast positive algorithm is required to process the time series containing the 32000 data points. The positive distribution of Figure No. 5 derived by using the fast Groutage algorithm requires 4 million bytes of RAM compared to 4 billion bytes using the original algorithm of Loughlin et al. This is an improvement by a factor of 1000. Likewise, from a computational standpoint, the Groutage algorithm requires 2 million integrations per iteration in the solution process for converging on an answer associated with the set of nonlinear integral equations. This compares to 2 billion integrations per integration for the original Loughlin et al. algorithm. This also is an improvement by a factor of 1000. The iteration process is required to solve the set of coupled nonlinear integral equations. Depending upon the desired accuracy for the solu-

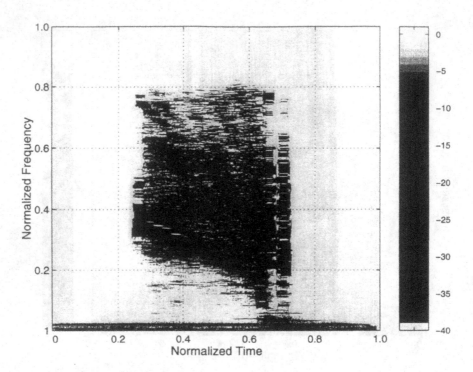

Figure 4. Narrow band spectrogram.

tion, numerous iterations are required until the solution converges to a pre specified error bound.

III. Conclusions

The digital time series representation of a non-stationary acoustic transient signal that emanates from a under water vehicle can contain tens of thousands to hundreds of thousands of data points. Not only must the processing method be able to handle the large number of data points in the signal, but also be such that an accurate joint time-frequency energy density can be obtained from the digital time series. The current method that is being used by the under water vehicle community, the one-third octave method, does not yield an accurate joint time-frequency energy density representation. Furthermore, spectrograms do not yield accurate energy density representations. The spectrogram is a function of the window length chosen to construct it. For each window length, a different spectrogram is generated, and therefore, a different joint time-frequency energy density representation. In essence, the problems that are associated with both the one-third octave method and the spectrogram are that neither technique satisfies the marginals, neither yield an accurate energy density, and both have poor resolution in time or in frequency, or in both time and

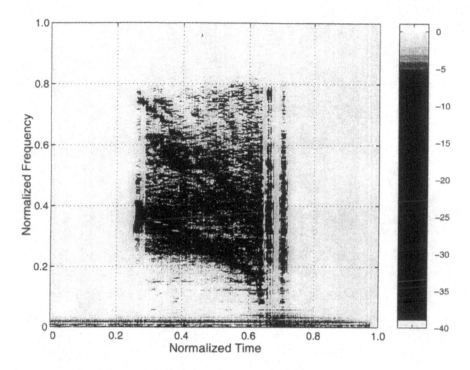

Figure 5. Positive distribution.

frequency simultaneously. The positive time-frequency distribution developed by Loughlin et al. alleviates the problems of both the one-third octave and spectrogram methods. The original method of Loughlin et al., however, is not computationally efficient when dealing with real world signals that contain tens of thousands of data points. The new algorithm of Groutage is computationally efficient for processing large time series records that contain up to hundreds of thousands of data points. The new algorithm when operating on a signal with 3200 data points runs thousands of times faster and requires thousands of times less memory than the Loughlin method.

References

1. L. Cohen, "Generalized Phase-Space Distribution Function," *J. Math. Phys.*, vol. 7 no. 5, 1966, pp. 781–786.
2. L. Cohen, and T. Posch, "Positive Time-Frequency Distribution Functions," *IEEE Trans. Acoust, Speech Signal Processing*, vol. ASSP-33, no. 1, 1985, pp. 31–38.
3. L. Cohen, and Y. Zaparovanny, "Positive Quantum Joint Distributions," *J. Math. Phys.*, vol. 21, no. 4, 1980, pp. 794–796.
4. P. Loughlin, J. Pitton, and L. Atlas, "Construction of Positive TFDs," *IEEE Trans. Sig. Proc.*, vol. 42, no. 10, 1994, pp. 2697–2705.
5. D. Groutage, "Fast Algorithm for Computing Minimum Cross-Entropy Positive Time-Frequency Distribution," *IEEE Trans. Sig. Proc.*, vol. 45, no. 8, August 1997.

Multidimensional Systems and Signal Processing, 9, 391–398 (1998)
© 1998 Kluwer Academic Publishers, Boston. Manufactured in The Netherlands.

Multicomponent Multidimensional Signals

JOSEPH P. HAVLICEK*   joebob@tobasco.ecn.ou.edu
School of Electrical and Computer Engineering, The University of Oklahoma, Norman, OK 73019-1023

DAVID S. HARDING   dave@vision.ece.utexas.edu
Center for Vision and Image Sciences, The University of Texas, Austin, TX 78712-1084

ALAN C. BOVIK   bovik@orion.ece.utexas.edu
Center for Vision and Image Sciences, The University of Texas, Austin, TX 78712-1084

Received August 12, 1997; Accepted January 6, 1998

Abstract. In this brief paper, we extend the notion of multicomponent signal into multiple dimensions. A definition for multidimensional instantaneous bandwidth is presented and used to develop criteria for determining the multicomponent nature of a signal. We demonstrate application of the criteria by testing the validity of a multicomponent interpretation for a complicated nonstationary texture image.

Key Words: Multicomponent signals, instantaneous frequency, instantaneous bandwidth, AM-FM models

1. Introduction

Signal descriptions that are inherently capable of capturing nonstationary structure are of great practical interest in an increasing variety of signal processing applications. For many signals, representation in terms of instantaneously varying quantities such as amplitude and frequency are fundamental as well as intuitively appealing. For example, a pure FM chirp is most naturally described as a constant-modulus exponential with linearly increasing frequency. More generally, a nonstationary signal $t : \mathbb{R} \to \mathbb{C}$ may be modeled by the joint amplitude-frequency modulated *AM-FM function*

$$t(x) = a(x)e^{j\varphi(x)}, \tag{1}$$

where $a(x)$ and $\varphi(x)$ are unique; $a(x)$ is referred to as the *instantaneous amplitude*, or *amplitude modulation function* of $t(x)$, whereas $\varphi'(x)$ is known as the *instantaneous frequency*, or *frequency modulation function*. A real signal $s : \mathbb{R} \to \mathbb{R}$ may be analyzed against the model (1) using the unique complex extension $t(x) = s(x) + j\mathcal{H}[s(x)]$, known as the *analytic signal* [6,11], where $\mathcal{H}$ indicates the Hilbert transform. With the analytic signal, the amplitude and frequency of a real-valued signal are unambiguously defined in a way that establishes attractive fundamental relationships between the instantaneous frequency and Fourier spectrum of the signal [1,3,5,6,10,11].

The model (1) does not deliver an intuitively satisfying interpretation for *all* signals, however. Consider the signal $t(x) = a_1 e^{j\omega_1 x} + a_2 e^{j\omega_2 x}$ [3,10]. Intuitively, $t(x)$ is the

---
* This research was supported in part by the Army Research Office under contract DAAH 049510494 and by the Air Force Office of Scientific Research, Air Force Systems Command, USAF, under grant number F49620-93-1-0307.

sum of two components each having constant amplitude and frequency. The interpretation delivered by (1) is

$$a(x) = \sqrt{a_1^2 + a_2^2 + 2a_1 a_2 \cos[(w_2 - w_1)x]} \tag{2}$$

and

$$\varphi'(x) = \tfrac{1}{2}(w_2 + w_1) + \tfrac{1}{2}(w_2 - w_1)\frac{a_2^2 - a_1^2}{a^2(x)}, \tag{3}$$

both of which oscillate for all nontrivial choices of the parameters. Indeed, certain signals are *inherently multipartite* in character and are better interpreted as a *sum* of components that each take the form (1).

Cohen and Lee have developed the notion of *multicomponent signal* in 1D [2–5]. They introduced the *instantaneous bandwidth*, which for $t(x)$ in (1) is defined by $B(x) = |a'(x)/a(x)|$. Within the context of certain quadratic time-frequency distributions, $B^2(x)$ admits a rigorous interpretation as the conditional instantaneous spread of frequency about $\varphi'(x)$. Cohen and Lee consider a signal to be multicomponent if there exists a decomposition into components of the form (1) such that the instantaneous bandwidth of each component is smaller than the instantaneous bandwidth of the composite signal and such that the frequency separation between components is large compared to their instantaneous bandwidths. In this brief paper, we discuss the extension of this notion of multicomponent signal into multiple dimensions.

2. Multicomponent Multidimensional Signals

For a multidimensional signal $t : \mathbb{R}^n \to \mathbb{C}$ modeled by the multicomponent AM-FM function

$$t(\mathbf{x}) = \sum_{i=1}^{K} a_i(\mathbf{x}) \exp[j\varphi_i(\mathbf{x})] = \sum_{i=1}^{K} t_i(\mathbf{x}), \tag{4}$$

we define the instantaneous bandwidth of component $t_i(\mathbf{x})$ by [7,8]

$$B_i(\mathbf{x}) = \left| \frac{\nabla a_i(\mathbf{x})}{a_i(\mathbf{x})} \right| = \left| \mathrm{Im}\left[\frac{\nabla t_i(\mathbf{x})}{j t_i(\mathbf{x})} \right] \right|. \tag{5}$$

The magnitudes of the individual components of the vector $\nabla a_i(\mathbf{x})/a_i(\mathbf{x})$ are analogous to the 1D instantaneous bandwidth, and describe the local spread of frequencies in each dimension. $B_i(\mathbf{x})$ in (5) quantifies the spread simultaneously in *all* dimensions. The instantaneous bandwidth for the composite signal $t(\mathbf{x})$ is obtained by taking $K = 1$ in (4) and applying (5). A real-valued signal may be analyzed against the model (4) by applying the directional multidimensional Hilbert transform described in [9].

We consider that $t(\mathbf{x})$ is multicomponent on a region $S \subset \mathbb{R}^n$ if a decomposition of the form (4) exists over S with $K > 1$ such that two conditions are satisfied. First,

Figure 1. Reptile texture image.

the instantaneous bandwidth of each component must be appreciably smaller than the instantaneous bandwidth of $t(\mathbf{x})$. Second, the frequency separation between any pair of components in the multicomponent interpretation must be large compared to the component instantaneous bandwidths on a pointwise basis. Thus, for each i and each j in $[1, K]$ we require that

$$B_i(\mathbf{x}), B_j(\mathbf{x}) \ll \left| \nabla \varphi_i(\mathbf{x}) - \nabla \varphi_j(\mathbf{x}) \right|. \tag{6}$$

Note that, as in 1D, this notion of multicomponent signal implies that $t(\mathbf{x})$ may generally be multicomponent in certain regions and not in others.

3. Example

The nonstationary, multipartite texture image *Reptile* is shown in Fig. 1. A six-component interpretation of the image is given in Fig. 2, where components one through six appear in parts (a)–(f) respectively. These components were extracted using the multicomponent AM-FM demodulation techniques described in [8]. In Fig. 2, each component has been independently scaled for display. Note that all of the components exhibit significant nonstationary structure manifest as spatially varying amplitude and frequency modulations.

The amplitude of the composite image computed using the multidimensional Hilbert transform is given in Fig. 3(a). Fig. 3(b) gives the instantaneous bandwidth for the composite image, which has a mean value of approximately 0.35 and lies between zero and 24.0. For comparison, the amplitude and instantaneous bandwidth of component two are shown in Fig. 4(a) and (b), respectively. The instantaneous bandwidth of component two lies between zero and 15×10^{-3}. Its mean value is approximately 5×10^{-3}.

A histogram of $B(\mathbf{x})$ for the composite *Reptile* image appears in Fig. 5(a), while instantaneous bandwidth histograms for components one through six are given in Fig. 5(b)–(g).

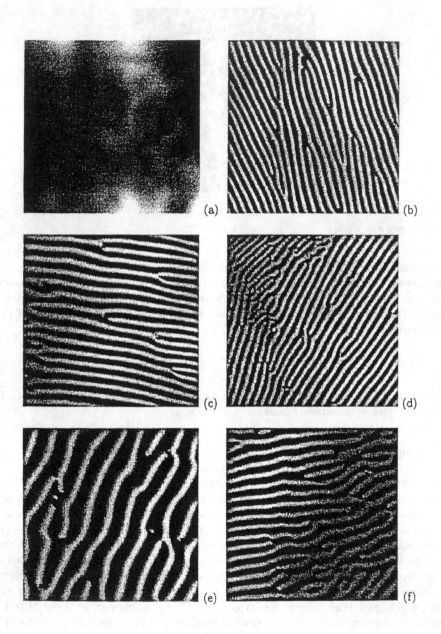

Figure 2. Six-component interpretation of *Reptile* image. (a) Component one. (b) Component two. (c) Component three. (d) Component four. (e) Component five. (f) Component six.

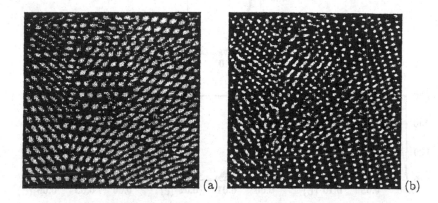

Figure 3. Amplitude and instantaneous bandwidth of composite *Reptile* image. (a) Computed amplitude modulation function $a(\mathbf{x})$. (b) Instantaneous bandwidth $B(\mathbf{x})$.

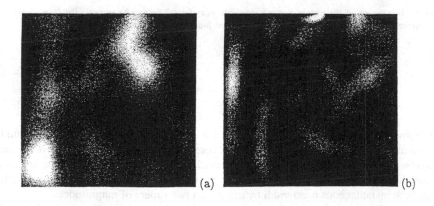

Figure 4. Amplitude and instantaneous bandwidth of component two. (a) Computed amplitude modulation function $a_2(\mathbf{x})$. (b) Instantaneous bandwidth $B_2(\mathbf{x})$.

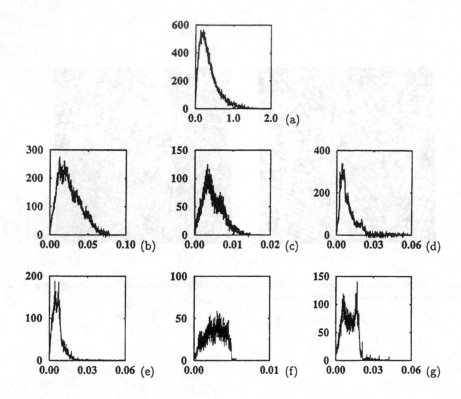

Figure 5. Histograms of the instantaneous bandwidth $B(\mathbf{x})$ for (a) composite image. (b) Component one. (c) Component two. (d) Component three. (e) Component four. (f) Component five. (g) Component six.

Each histogram in Fig. 5(a)–(g) depicts the same number of data points. The areas under the various curves appear to be different because different bin sizes were used for each histogram in order to accurately reflect the spread of values assumed by the instantaneous bandwidth. Note that, on average, the decomposition of this image into components has reduced the instantaneous bandwidth by more than two orders of magnitude.

The ratio of $B_2(\mathbf{x})$ to the quantity on the right side of (6) is histogrammed in Fig. 6 for $i = 2$ and $j = 1, 3, \ldots, 6$. Thus, small abscissa values in these histograms indicate points where the frequency separation between components is large compared to the instantaneous bandwidth of component two. Collectively, the histograms in Fig. 5 and Fig. 6 strongly indicate that the *Reptile* image is indeed multicomponent and that the multicomponent interpretation depicted in Fig. 2 is a valid one.

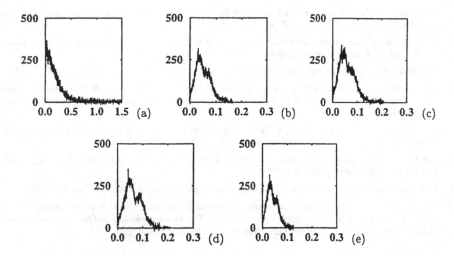

Figure 6. Histograms of (a) the ratio of frequency separation between components one and two to instantaneous bandwidth of component two. (b) the ratio of frequency separation between components two and three to instantaneous bandwidth of component two. (c) the ratio of frequency separation between components two and four to instantaneous bandwidth of component two. (d) the ratio of frequency separation between components two and five to instantaneous bandwidth of component two. (e) the ratio of frequency separation between components two and six to instantaneous bandwidth of component two.

4. Discussion

The two conditions discussed in Section 2 imply that a multidimensional signal is multicomponent if it can be decomposed into a sum of components that are well delineated in instantaneous frequency and that are tightly concentrated on a local basis in the time-frequency or space/spatial frequency hyperplanes. Decompositions that satisfy these conditions generally tend to be physically meaningful and intuitively satisfying. This notion of multicomponent signal does not, however, suggest a procedure for decomposing a multipartite multidimensional signal into components. The computation of valid multicomponent interpretations for complicated natural images and video is extremely difficult in general and remains an active area of research.

References

1. L. Cohen, "Distributions Concentrated Along the Instantaneous Frequency," *SPIE Adv. Signal Proc. Alg., Architectures, Impl.*, vol. 1348, 1990, pp. 149–157.

2. L. Cohen, "What is a Multicomponent Signal?" *Proc. IEEE Int'l. Conf. Acoust., Speech, Signal Proc.*, vol. V, pp. 113–116, San Francisco, CA, March 1992.

3. L. Cohen, *Time-Frequency Analysis*, Englewood Cliffs, NJ: Prentice Hall, 1995.

4. L. Cohen and C. Lee, "Instantaneous Frequency, Its Standard Deviation and Multicomponent Signals," *SPIE Adv. Alg. Architectures Signal Proc. III*, vol. 975, 1988, pp. 186–208.

5. L. Cohen and C. Lee, "Instantaneous Bandwidth," B. Boashash, editor, *Time-Frequency Signal Analysis*, pp. 98–117. Melbourne: Longman Cheshire, 1992.

6. D. Gabor, "Theory of Communication," *J. Inst. Elect. Eng. London*, vol. 93, no. III, 1946, pp. 429–457.

7. J. P. Havlicek, A. C. Bovik, and P. Maragos, "Modulation Models for Image Processing and Wavelet-Based Image Demodulation," *Proc. 26th IEEE Asilomar Conf. Signals, Syst., Comput.*, pp. 805–810, Pacific Grove, CA, October 26–28, 1992.

8. J. P. Havlicek, D. S. Harding, and A. C. Bovik, "The Multi-Component AM-FM Image Representation," *IEEE Trans. Image Proc.*, vol. 5, no. 6, 1996, pp. 1094–1100.

9. J. P. Havlicek, J. W. Havlicek, and A. C. Bovik, "The Analytic Image," *Proc. IEEE Int'l. Conf. Image Proc.*, Santa Barbara, CA, October 26–29, 1997.

10. L. Mandel, "Interpretation of Instantaneous Frequencies," *Am. J. Phys.*, vol. 42, 1974, pp. 840–846.

11. J. Ville, "Théorie et Applications de la Notation de Signal Analytique," *Cables et Transmission*, vol. 2A, 1948, pp. 61–74. Translated from the French in I. Selin, "Theory and applications of the notion of complex signal," Tech. Rept. T-92, The RAND Corporation, Santa Monica, CA, August, 1958.

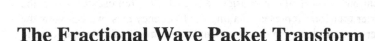
Multidimensional Systems and Signal Processing, 9, 399–402 (1998)
© 1998 Kluwer Academic Publishers, Boston. Manufactured in The Netherlands.

The Fractional Wave Packet Transform

YING HUANG*   yhuang@afit.af.mil
Dept. of Electrical & Computer Engineering, Air Force Institute of Technology, AFIT/ENG, 2950 P Street, Wright-Patterson AFB, OH 45433-7765

BRUCE SUTER
Dept. of Electrical & Computer Engineering, Air Force Institute of Technology, AFIT/ENG, 2950 P Street, Wright-Patterson AFB, OH 45433-7765, USA

Received July 16, 1996; Revised February 1, 1997

Abstract. We introduce the concept of the Fractional Wave Packet Transform(FRWPT), based on the idea of the Fractional Fourier Transform(FRFT) and Wave Packet Transform(WPT). We show a version of the resolution of the identity and some properties of FRWPT connected with those of FRFT and WPT.

Key Words: Fractional Fourier Transform, wavelet, Wave Packet Transform

1. Introduction

The short-time Fourier transform (STFT), or windowed Fourier transform, is the most widely used method in signal processing for studying nonstationary signals [1]. The continuous STFT of a signal $f(t) \in L^2(\mathbb{R})$ is defined as

$$\frac{1}{\sqrt{2\pi}} \int_{-\infty}^{+\infty} e^{-jut} g(t - \tau) f(t) dt,$$

where $g(t)$ is the window function. The wavelet transform [2] gives a similar time-frequency description. With an analyzing wavelet ψ, the continuous wavelet transform (CWT) is then

$$\frac{1}{\sqrt{a}} \int_{-\infty}^{+\infty} \psi\left(\frac{t - b}{a}\right) f(t) dt.$$

Here $a > 0$ and ψ is normalized such that the L^2 norm $\|\psi\| = 1$. The wave packet transform (WPT) (see, for example, [3,4,5,6]) can be viewed as the "combination" of the STFT and CWT:

$$\frac{1}{\sqrt{2\pi a}} \int_{-\infty}^{+\infty} e^{-jut} \psi\left(\frac{t - b}{a}\right) f(t) dt.$$

In other words, the WPT is the Fourier transform of a signal windowed with a wavelet that is dilated by a and translated by b.

Fractional Fourier transform (FRFT) [7] is the generalization of the conventional Fourier transform. Studies of FRFT have already led to signal processing applications(See, e.g.,

---
* This work was supported in part by the Air Force Office of Scientific Research under Grants AFOSR-PO-94-0004, AFOSR-PO-95-0004, AFOSR-616-95-0022, and AFOSR-616-96-0054.

[8,10,9,11]). As mentioned in [7], if we consider a signal $f(t)$ represented along the time axis and its Fourier transform represented along the frequency axis, we can view the Fourier transform operator as a change in the representation of the signal corresponding to a counterclockwise axis rotation of $\pi/2$ rad. The FRFT then corresponds to a rotation by an angle that is not necessarily a multiple of $\pi/2$. It is defined via

$$\int_{-\infty}^{+\infty} \mathcal{K}_\alpha(t, u) f(t) dt,$$

where the transformation kernel $\mathcal{K}_\alpha$ is defined as the following:

$$\mathcal{K}_\alpha(t, u) = \begin{cases} \sqrt{\frac{1-j\cot\alpha}{2\pi}} e^{j\frac{t^2+u^2}{2}\cot\alpha - jut\csc\alpha} \\ \quad \text{if } \alpha \text{ is not a multiple of } \pi \\ \delta(t-u) \\ \quad \text{if } \alpha \text{ is a multiple of } 2\pi \\ \delta(t+u) \\ \quad \text{if } \alpha + \pi \text{ is a multiple of } 2\pi. \end{cases}$$

Note that when $\alpha = \pi/2$, $\mathcal{K}_\alpha$ is the kernel of the Fourier transform.

In this paper, we introduce the fractional wave packet transform (FRWPT) $(\mathcal{F}_\alpha f)(u, a, b)$ for a given a signal $f(t)$ as following:

$$(\mathcal{F}_\alpha f)(u, a, b) = \frac{1}{\sqrt{a}} \int_{-\infty}^{+\infty} \mathcal{K}_\alpha \psi\left(\frac{t-b}{a}\right) f(t) dt.$$

More precisely, for angles that are not multiples of π, the fractional wave packet transform is:

$$(\mathcal{F}_\alpha f)(u, a, b) = \sqrt{\frac{1-j\cot\alpha}{2\pi}} e^{j\frac{u^2}{2}\cot\alpha} \int_{-\infty}^{+\infty} \psi\left(\frac{t-b}{a}\right) f(t) e^{j\frac{t^2}{2}\cot\alpha} e^{jut\csc\alpha} dt.$$

Note that the FRWPT $\mathcal{F}_\alpha$ is a function of time, frequency and scale. Note also that for $\alpha = \pi/2$, the fractional wave packet transform coincides with the wave packet transform. Following the steps as explained in [7], it is easy to see that the computation of the FRWPT corresponds to the following steps:

(1) a product by a wavelet

(2) a product by a chirp

(3) a Fourier transform

(4) another product by a chirp

(5) a product by a complex amplitude factor.

68

2. WPT, FRFT and FRWPT

The $\mathcal{K}_\alpha$ also has the following properties:

$$\mathcal{K}_\alpha(t, u) = \mathcal{K}_\alpha(u, t)$$
$$\mathcal{K}_\alpha^*(t, u) = \mathcal{K}_{-\alpha}(t, u)$$
$$\int_{-\infty}^{+\infty} \mathcal{K}_\alpha(t, u)\mathcal{K}_\beta(u, z)du = \mathcal{K}_{\alpha+\beta}(t, z)$$
$$\int_{-\infty}^{+\infty} \mathcal{K}_\alpha(t, u)\mathcal{K}_\alpha^*(t, u')dt = \delta(u - u').$$

The last property is most important since it means that the kernel functions $\mathcal{K}_\alpha(t, u)$, taken as functions of t with parameter u, form an othonormal set. This property can be used to prove a version of resolution of the identity as showed below, which is similar to that of Wavelet transform and the STFT.

If we define $\psi_a(t) = \frac{1}{\sqrt{a}}\psi\left(\frac{t}{a}\right)$, then $\|\psi_a\| = \|\psi\| = 1$. Recall the definition of the WPT, we can now view WPT as the STFT with window function ψ_a. Therefore we can show a version of the Resolution of the Identity for FRWPT as the following:

$$\int_{-\infty}^{+\infty}\int_{-\infty}^{+\infty} dudb(\mathcal{F}_\alpha f)(u, a, b)(\mathcal{F}_\alpha g)^*(u, a, h)$$
$$= \int\int du\, db \int \mathcal{K}_\alpha(u, t)\psi_a(t - b)f(t)dt \times \int \mathcal{K}_\alpha(u, t')\psi_a(t' - b)g(t')dt'$$
$$= \int db \int\int \delta(t - t')\psi_a(t - b)\psi_a^*(t' - b) \times f(t)g^*(t')dtdt'$$
$$= \int\int dbdt|\psi_a(t - b)|^2 f(t)g^*(t)$$
$$= \|\psi_a\| \langle f, g \rangle = \langle f, g \rangle.$$

As a consequence, we can derive the energy-preserving property of FRWPT:

$$\int_{-\infty}^{+\infty}\int_{-\infty}^{+\infty} |\mathcal{F}_\alpha(a, b, u)|^2 dudb = \int_{-\infty}^{+\infty} |f(t)|^2 dt.$$

Remark: Using the energy-preserving property of the FRFT, we also have

$$\int_{-\infty}^{+\infty} |\mathcal{F}_\alpha(a, b, u)|^2 du = \int_{-\infty}^{+\infty} |\psi_a(t - b)f(t)|^2 dt.$$

In [3], the Wavepacketgram is defined as the modulus square of the WPT and it is proved that the Wavepacketgram is a memeber of the Cohen class time-frequency distributions where the kernel is a scale dependent ambiguity function. Similarly, we can define the fractional Wavepacketgram of the signal $f(t)$, with angle α, as $|\mathcal{F}_\alpha(a, b, u)|^2$.

3. Conclusion

We have introduced the concept of the Fractional Wave Packet Transform(FRWPT), combining the idea of the Fractional Fourier Transform(FRFT) and Wave Packet Transform(WPT). We have obtained a version of the resolution of the identity and some properties of FRWPT connected with those of FRFR and WPT. Further study can be done on the relationships of the FRWPT with other time-frequency representations such as Wigner distribution, the ambiguity function and the spectrogram.

References

1. L. Cohen, *Time-Frequency Analysis*, Englewood Cliffs, New Jersey: Prentice Hall, 1995.

2. I. Daubechies, "The Wavelet Transform, Time-Frequency Localization and Signal Analysis," *IEEE Trans. Inform. Theory*, vol. 36, no. 11, 1990, pp.961–1005.

3. T. Posch, "The Wave Packet Transform (WPT) as Applied to Signal Processing," *Proc. IEEE-SP Intl. Sym. Time-Frequency and Time-Scale Analysis*, Victoria, BC, Canada, October, 1992, pp. 143–146.

4. B. Torresani, "Wavelets Associated with Representations of the Affine Weyl-Heisenberg Group," *J. Math. Phys.*, vol. 32, no. 5, 1991.

5. B. Torresani, "Time-Frequency Representations: Wavelet Packets and Optimal Decomposition," *Ann. Inst. Henri Poincaré*, vol. 56, no. 2, 1992, pp. 215–234.

6. P. Guillemain, and R. Kronland-Martinet, "Characterization of Acoustic Signals Through Continuous Linear Time-Frequency Representations," *Proceedings of the IEEE*, vol. 84, no. 4, 1996.

7. L. Almeida, "The Fractional Fourier Transform and Time-Frequency Representations," *IEEE Trans. Signal Proc*, vol. 42, no. 11, 1994.

8. A. Lohmann, "Image Rotation, Wigner Rotation and Fractional Fourier Transform," *J. Opt. Soc. Amer. A*, vol. 10, 1993, pp. 2181–2186.

9. H. Ozaktas, and D. Mendlovic, "Fourier Transform of Fractional Order and Their Optical Interpretation," *Optics Comm.*, vol. 101, 1993, pp. 163–169,

10. A. Lohmann, and B. Soffer, "Relationships Between the Radon-Wigner and Fractional Fourier Transforms," *J. Opt. Soc. Amer. A*, vol. 11, 1994, pp. 1798–1801.

11. S.-Y. Lee and H. Szu, "Fractional Fourier Transforms, Wavelet Transforms, and Adaptive Neural Networks," *Opt. Eng.*, vol. 33, 1994, pp. 1159–1161.

Multidimensional Systems and Signal Processing, 9, 403–409 (1998)

Statistical Performance of a TFD based Detection Scheme

OWEN P. KENNY*
Signal Analysis Discipline, Communication Division, Defence Science and Technology Organisation, PO Box 1500 Salisbury SA 5108, AUSTRALIA

LANGFORD B. WHITE
Signal Analysis Discipline, Communication Division, Defence Science and Technology Organisation, PO Box 1500 Salisbury SA 5108, AUSTRALIA

Received November 15, 1996; Revised April 5, 1997

Abstract. This paper proposes a robust signal detection procedure based on generalised patterns in time-frequency space. The paper derives a non-linear time domain correlation process which is equivalent to a time-frequency based detection scheme. The performance of this robust detection scheme in the additive white Gaussian noise case is evaluated with the aid of *receiver operating curves*. Some examples are given which show that the proposed procedure under nominal conditions does not perform as well as standard detection scheme (ie matched filter) but gives better performance when the measured signal deviates from its nominal model.

Key Words: time-frequency distributions, robust detection

1. Introduction

The signal processing requirement concerned within this paper is the robust detection of signals. With traditional detection processes all the information about the process is assumed to be known, ie. both the deterministic and random nature of the process. In such cases, detection schemes optimise the detection probability or the signal to noise ratio output resulting in the well known matched filter. Matched filters, however, depend on the detailed knowledge of the exact signal model and the noise statistics. It is well known that the performance of the matched filter may degrade dramatically when the signal deviates from the proposed model [1]. For example, the output of the matched filter for the detection of a sinusoidal function at a nominal frequency degrades according to a sinc function as the signal frequency deviates from the nominal frequency. This highlights the need for detection algorithms to be inherently insensitive to signal model assumptions, particularly in a practical detection situation.

The paper first outlines a general description of a detection scheme based on the Wigner distributions (WD) [2]. Such a scheme is parameterised by a user-specified time-frequency pattern function which is chosen to represent the properties of the class of signals one is seeking to detect. An eigenvalue decomposition of this pattern function, together with Moyal's identity, is used to obtain a time domain representation for the detector. The core contribution of the work is the derivation of the analytic performance equations for the additive white Gaussian noise case. Finally, we present some illustrative examples.

---

* Appeared in part in *IEEE ICASSP 1995*, Detroit, USA.

2. Problem Definition and the Proposed Detector

It is well known that for any finite energy signals $x(.)$ and $y(.)$, the inner product of a signal is given by

$$\langle x, y \rangle = \int_{-\infty}^{+\infty} x(t)y^*(t)dt \tag{1}$$

and Schwartz's inequality

$$|\langle x, y \rangle| \leq \|x\| \|y\| \tag{2}$$

holds. Both (1) and (2) form the basis of time domain matched filtering, and can be used to classify a measured signal as one of the specified nominal functions $y_1 \ldots y_N$. The decision rule in traditional classification procedures is to choose the signal which maximises the output of (2) over the various test functions $y_i(.)$.

It can be shown that the inner product of the WD of the measured signal $y(.)$ with the WD of a known signal $x(.)$ is related to the inner product defined in (2) [2] via

$$\int_{R^2} W_x(t, f) W_y(t, f) dt \, df = |\langle x, y \rangle|^2 \tag{3}$$

where by definition, the WD is

$$W_x(t, f) = \int_{-\infty}^{+\infty} x \left(t + \frac{\tau}{2} \right) x^* \left(t - \frac{\tau}{2} \right) e^{-i2\pi f \tau} d\tau. \tag{4}$$

Thus, the WD can form the basis of a correlation classifier; however, it appears necessary that the WD of the signal under analysis be calculated. Since by Moyal's identity, the two formulations for the correlation statistic are equal then one might asked the question—"why use time-frequency distributions at all to formulate the correlation statistic?" Replacement of the WD by a more general form of time-frequency pattern function provides the answer to this question. This allows much greater flexibility in the classification processs, as representation of a broader test class of signals is possible. We now define a new correlation statistic by

$$\eta_j(x) = \langle W_x, \psi_j \rangle, \quad j = 1, \ldots, N, \tag{5}$$

where ψ_j denote representative time-frequency patterns which specify a given class of signals to be detected. It can be shown [3] that if ψ_j is not a WD, then (5) cannot be represented as a single one dimensional correlation function like (2).

Generally, the windowing function ψ_j has Hermitian symmetry so that the bilinear time-frequency representation is real. We also assume here that ψ_j is chosen such that (5) is non-negative although this is not restrictive. We show in [3] that η_j is identical to

$$\eta_j(x) = \sum_{n=1}^{\infty} \lambda_n^{(j)} |\langle x, v_n^{(j)} \rangle|^2 \tag{6}$$

where the $\lambda_n^{(j)} \geq 0$ and $v_n^{(j)}$ are the eigenvalues and eigenfunctions of a compact non-negative self-adjoint operator specified by ψ_j. Effective truncation of (6) to N terms can be achieved by retaining those terms with the N-th largest eigenvalues. The reader is referred to [3] for details.

3. Derivation of Analytic Receiver Operating Characteristics

In this section, we derive the set of analytic equations that describe the detector performance known as the receiver operating characteristics (ROC). Before deriving the required cumulative distribution function (cdf) for the non-linear detection statistic, some assumptions are made about the processes involved. Let the received signal be given by $y(t) = x(t) + n(t)$ where $x(.)$ is a deterministic signal and $n(.)$ denotes a white Gaussian noise process with zero mean and variance of σ^2. Let $\chi(.)$ and $\zeta(.)$ denote the projections of $x(.)$ and $n(.)$ respectively onto $Sp\{\phi_1, \ldots, \phi_N\}$ (dropping the subscript j denoting the pattern class chosen to test) with Fourier coefficients χ_n and ζ_n respectively then,

$$\eta(y) = \sum_{i=1}^{N} \lambda_i \, |\langle (x + n), \phi_i \rangle|^2 = \sum_{i=1}^{N} \lambda_I (\chi_i + \zeta_i)^2 = \sum_{i=1}^{N} \alpha_i (\delta_i + Z_i)^2 \tag{7}$$

where $\alpha_i = \sigma^2 \lambda_i$, $\delta_i = \chi_i/\sigma$, and $Z_i = \zeta_i/\sigma \sim N(0,1)$. Our aim is to determine the cdf $F_\eta(\alpha; \delta; y)$ of the detection statistic (7) and thus the probability of detection and of false detection. The probability of detection for a given threshold value ε is $P_d(\varepsilon) = 1 - F_\eta(\alpha; \delta; \varepsilon)$ whilst the false detection probability is $P_f(\varepsilon) = 1 - F_\eta(\alpha; \delta; \varepsilon)$. These equations specify the receiver operating curve which completely characterises the performance of the robust detection algorithm. Equation (7) represents a quadratic form in uncorrelated normalised Gaussian variables Z_n. The analytic expression for the probability of a such a form is not trivial but can be found in [4]. The first step is to obtain the characteristic function of (7):

$$L_\eta(s) = \exp\left\{ -\sum_{i=1}^{N} \frac{\delta_i^2 \alpha_i s}{(1 + 2\alpha_i \sigma^2 s)} \right\} \prod_{j=1}^{N} (1 + 2s\alpha_i \sigma^2)^{-\frac{1}{2}}. \tag{8}$$

We then may obtain an expansion of the cdf of the form,

$$F_\eta(\alpha; \delta; y) = \sum_{k=0}^{\infty} a_k F(n + 2k; y/\beta) \tag{9}$$

where $\beta > 0$ is arbitrary (this is chosen for reasons related to rapidity of convergence [4]) and $F(.)$ denotes the cdf for a non-central χ^2 variable. The coefficients a_k are determined from the recurrence relations

$$a_0 = \prod_{j=1}^{N} \left(\frac{\beta}{\alpha_j} \right)^{1/2} ; a_k = \frac{1}{k} \sum_{j=0}^{k-1} b_{k-r} a_k \quad k \geq 1$$

$$b_1 = \frac{1}{2} \sum_{j=1}^{N} (1 - \delta_j^2) \gamma_j; \ \gamma_j = 1 - \beta/\alpha_j \tag{10}$$

$$b_k = \frac{k}{2} \sum_{j=1}^{N} \delta_j^2 \gamma_j^{k-1} + \frac{1}{2} \sum_{j=1}^{N} (1 - k\delta_j^2) \gamma_j^k \ \ k \geq 2.$$

The example which follows in the next section shows the advantages and the cost incurred in using a robust matched filter.

4. An Example—Robust Detection of a Linear FM Signal with Unknown Chirp Rate

In this section an example is given to demonstrate the use of robust detection of signals. The robust detection scheme is compared to that of a matched filter. So that a fair comparison is made between these detection approaches the outputs of the matched filter and the robust detection are normalised so that each has the same noise power output. The signal under detection is a linear FM signal which has the form $x(t) = sin(2\pi f_o t + \alpha t^2)$, where f_o is the initial frequency and α denotes the chirp rate of the signal. Here the exact value of the chirp rate is not known, but it is assumed that lies within the interval $[\alpha_0, \alpha_1]$.

The region in which the signal energy is concentrated in the TF plane is shown in Figure 1. A matched filter was designed for the case where the value of the chirp rate was α_o. Figure 2 shows the output of the matched filter and the robust detection. Here as the signal's chirp rate increases it can be seen that the performance of the matched filter degrades, whereas the robust output still gave reasonable results. To further show the advantage of the robust detector the ROC where obtained for this example. As before the ROC where tested against the Monte Carlo simulations to verify the validity of the ROC. The results are shown in figure 3 indicating the analytic approach agreed with the simulated ROC. The nominal chirp rate in the experiment was 0.25 Hz/sec and the chirp rate was increased in steps of 0.005 until the chirp rate was 0.27 Hz/sec. The effect of the chirp rate is shown in figure 4 where it can be seen that the performance of the matched filter degraded as the chirp rate increased the curves approached the diagonal. The ROC for the robust detector plotted on the same graph shows that for all the variations in the chirp rate of the signal the pperformance of the robust detector stayed the same. What can be seen in both these examples is that the robust detector does not perform as well as the matched filter under matched conditions, but as the parameters of the signal under detection change the output of the matched filter degrades. The robust detector on the other hand can be designed to be invariant to these changes.

Conclusions

This paper proposed a robust detection statistic for signals whose parameters are uncertain. Standard detection schemes generally use time domain correlation which can be related to correlation based on the Wigner distribution (WD) by Moyal's identity. This paper showed

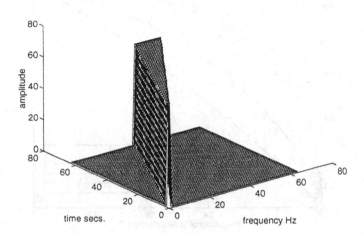

Figure 1. TF Pattern for example.

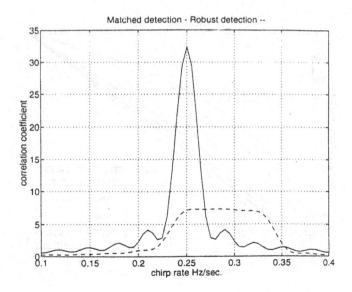

Figure 2. Detector outputs.

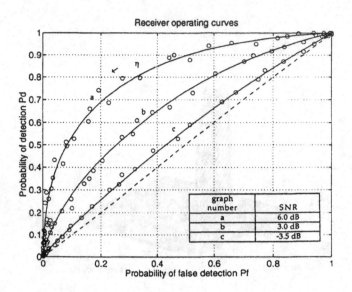

Figure 3. ROC curves—example.

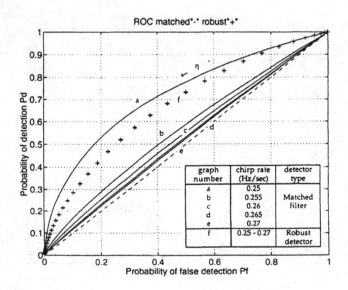

Figure 4. Performance of detector.

that a more robust detection scheme can be achieved by replacing the WD by generalised patterns in time-frequency space.

A generalised time-frequency pattern, that defines a class of detectable signals, was decomposed into a set of orthonormal eigenfunction and so by considering only the dominant eigenvalues a reduced order correlation statistic was obtained. This resulting statistic can not be otherwise implemented by a corresponding linear process illustrating the important feature that components inherent in the robust detector interact with each other to produce the desired response. The performance of the robust detection statistic was also evaluated with the aid of *receiver operating characteristics* (ROC) which involved the analytic derivation of the probabilities of detection and false alarm based on non-central χ^2 expansion of a quadratic form.

To gauge the effectiveness of the proposed robust detection statistic, we considered an example of a linear FM signal whose chirp rate was uncertain. The ROC curves showed the robust detector under nominal conditions did not perform as well as standard detection scheme, as one would expect, but gave better performance when the parameters of the signals deviated away from its nominal conditions.

References

1. S. A. Kassam, and H. V. Poor, "Robust Techniques for Signal Processing," *Proceeding of the IEEE*, vol. 73, no. 3, 1985.

2. L. Cohen, "Time-Frequency Distributions—A Review," *Proc. IEEE*, vol. 77, no. 7, 1989, pp. 941–981.

3. O. P. Kenny, and L. B. White, "Robust Detection in Time-Frequency Space," *Signals Analysis Discipline Report*, SA/95/06U, 1995.

4. S. Kotz, N. L. Johnson, and D. W. Boyd, "Series Representation of Distributions of Quadratic Forms in Normal Variables II. Non-Central Case," *Ann. of Statistical Math.*, 1966.

Multidimensional Systems and Signal Processing, 9, 411–418 (1998)
© 1998 Kluwer Academic Publishers, Boston. Manufactured in The Netherlands.

Unique Positive FM-AM Decomposition of Signals

RAMDAS KUMARESAN   kumar@ele.uri.edu
Department of Electrical Engineering, University of Rhode Island, Kingston, RI-02881

ASHWIN RAO
Department of Electrical Engineering, University of Rhode Island, Kingston, RI-02881

Received April 10, 1997; Accepted January 6, 1998

Abstract. An analytic signal, observed over an interval of T seconds, is decomposed into a Minimum Phase (AM) and an All-Phase (FM) signal. The former's log-envelope and phase form a Hilbert transform pair while the latter has positive definite instantaneous frequency (IF). An AM-FM demodulator algorithm to achieve the above decomposition is proposed.

Key Words: AM-FM decomposition, linear prediction in spectral domain (LPSD), analytic signals

1. Introduction

The fundamental issues related to analytic signals were addressed by Gabor in 1946 [1], followed by Ville and others [2]. A signal is said to be analytic if its Fourier transform vanishes for either positive or negative frequencies. Such a representation permits an unambiguous characterization of a signal by its envelope and phase/frequency modulations. The phase-derivative (or IF) of an analytic signal has been extensively studied [2,3]. A comprehensive review of IF of monocomponent signals (defined by Cohen [2]) has been provided by Boashash [4], along with discussions on existing algorithms and applications. The general impression among researchers has been that IF is meaningful only for narrowband or monocomponent signals [2]. In 1966, Voelcker studied the IF of analytic signals in the context of unifying various modulation methods. His studies were based on a nonlinear representation of signals as product of elementary signals rather than sums [5]. For a periodic bandlimited signal, the product-expansion simply means representing the signal by factoring the periodic signal's Fourier series. Non-periodic band-limited signals can be similarly treated as products of elementary signals (Cauchy-Hadamard product) but require difficult mathematics associated with the so-called entire functions [5]. However even when signals are not periodic and non-stationary, in many practical applications it is reasonable to work with a short segment of the signal and consider periodic extensions of it, as is common in short-time spectral analysis. In this paper we model a T second segment of a signal by a pole-zero model and decompose it into minimum-phase (MinP) and an all-phase (AllP) component whose IF is positive. An algorithm for such a decomposition is proposed.

This research was supported by a grant from the National Science Foundation under grant #CCR-9804050.

2. Product Representation of Signals

Let $s(t)$ be a periodic signal, with period T, consisting of $M + 1$ complex sinewaves. Let $\Omega = 2\pi/T$ denote its fundamental angular frequency. Then

$$s(t) = \sum_{k=0}^{M} a_k e^{jk\Omega t}, \tag{1}$$

where a_k's are the complex amplitudes of the sinusoids; $a_0 \neq 0$ and $a_M \neq 0$. We may regard $s(t)$ as a polynomial of degree M in the complex variable $e^{j\Omega t}$. Also, we may factor this polynomial into its M factors and rewrite $s(t)$ as

$$s(t) = \underbrace{\prod_{i=1}^{P}(1 - p_i e^{j\Omega t})}_{s_{\text{MinP}}(t)} \underbrace{A_0 e^{j\omega_0 t} \prod_{i=1}^{Q}\left(1 - \frac{1}{q_i}e^{-j\Omega t}\right)}_{s_{\text{TMaxP}}(t)}. \tag{2}$$

$p_1, p_2, \ldots, p_P$, and $q_1, q_2, \ldots, q_Q$ denote the polynomial's roots; $p_i = |p_i|e^{j\theta_i}$, $q_i = |q_i|e^{j\phi_i}$, $\omega_0 = Q\Omega$ and $A_0 = a_0(\prod_{i=1}^{Q} -q_i)$. p_is denote roots inside the unit circle in the complex plane, q_is are on or outside the unit circle. The subscript 'TMaxP' indicates that $s_{\text{TMaxP}}(t)$ is a maximum phase (MaxP) signal that has been translated in frequency by ω_0.

The above expressions, representing a bandlimited periodic signal may be recognized as the counterpart of the frequency response of finite impulse response (FIR) filters in discrete-time systems theory [6]. More generally, if $s(t)$ consists of an infinite number of spectral lines, i.e., $S(\omega) = \sum_{k=0}^{k=\infty} a_k \delta(\omega - k\Omega)$, then analogous to an (infinite impulse response) IIR filter's frequency response, we can represent $s(t)$ over T secs to desired accuracy using sufficient number of poles and zeros as

$$s(t) = A_0 \underbrace{\frac{\prod_{i=1}^{P}(1 - p_i e^{j\Omega t})}{\prod_{i=1}^{U}(1 - u_i e^{j\Omega t})}}_{s_{\text{MinP}}(t)} e^{j\omega_0 t} \underbrace{\prod_{i=1}^{Q}\left(1 - \frac{1}{q_i}e^{-j\Omega t}\right)}_{s_{\text{TMaxP}}(t))}; \tag{3}$$

p_is and q_is correspond to zeros inside and outside the unit circle respectively. u_is correspond to the signal's poles.

3. Envelope and Phase Relationships

An elementary signal [5], $e(t)$, is defined as

$$e(t) = 1 - pe^{j\Omega t}, \tag{4}$$

where $p = |p|e^{j\theta}$. If $|p| < 1$ then $e(t)$ is called a MinP signal since no other signal with the same envelope has a smaller phase angle. Observe that $|e(t)| > 0$. Taking the natural logarithm of both sides and using the series expansion, $\ln(1 - y) = \sum_{k=1}^{\infty} \frac{-y^k}{k}$, we get

$$\ln(1 - pe^{j\Omega t}) = \sum_{k=1}^{\infty} \frac{-p^k e^{-jk\Omega t}}{k}. \tag{5}$$

After exponentiating both sides we get the following identity:

$$1 - pe^{j\Omega t} = \exp\left(\sum_{k=1}^{\infty} \frac{-|p|^k}{k} \cos(k\Omega t + k\theta) + j \sum_{k=1}^{\infty} \frac{-|p|^k}{k} \sin(k\Omega t + k\theta)\right). \tag{6}$$

From the above expression we note that for an elementary MinP signal, $e(t)$, the logarithm of its envelope and its phase angle are related through the Hilbert transform. Similarly, for an elementary MaxP signal $(1 - qe^{j\Omega t}$, where $q = |q|e^{j\phi}$ and $|q| > 1)$ we get the identity

$$1 - qe^{j\Omega t} = (-qe^{j\Omega t}) \exp\left(\sum_{k=1}^{\infty} \frac{-|1/q|^k}{k} \cos(k\Omega t + k\phi)\right.$$

$$\left. - j \sum_{k=1}^{\infty} \frac{-|1/q|^k}{k} \sin(k\Omega t + k\phi)\right). \tag{7}$$

The key difference between Eq. (7) and Eq. (6) is a sign change in their phase functions. Using the above identities in Eq. 2 we have

$$s_{\text{MinP}}(t) = e^{\alpha(t)+j\hat{\alpha}(t)} \quad \text{and} \tag{8}$$

$$s_{\text{TMaxP}}(t) = A_0 e^{\beta(t)+j\left(\omega_0 t - \hat{\beta}(t)\right)} \quad, \quad \text{where} \tag{9}$$

$$\alpha(t) = \sum_{k=1}^{\infty} \sum_{i=1}^{P} -\frac{|p_i|^k}{k} \cos(k\Omega t + k\theta_i) \quad \text{and} \tag{10}$$

$$\beta(t) = \sum_{k=1}^{\infty} \sum_{i=1}^{Q} -\frac{1/|q_i|^k}{k} \cos(k\Omega t + k\phi_i). \tag{11}$$

Thus $s(t)$ can be compactly represented as

$$s(t) = A_c e^{j\omega_c t} e^{\alpha(t)+j\hat{\alpha}(t)} e^{\beta(t)-j\hat{\beta}(t)}, \tag{12}$$

where A_c corresponds to the overall amplitude of the signal and ω_c denotes its 'carrier' frequency. ω_c is equal to $\omega_0 = Q\Omega$ plus any arbitrary frequency translation that the signal $s(t)$ may have been subjected to. The log-envelope, $\alpha(t) + \beta(t) + \ln A_c$ and phase, $\omega_0 t + \hat{\alpha}(t) - \hat{\beta}(t)$, of $s(t)$, based on the above expressions, are clearly not band-limited functions. It can be shown that $|s(t)|^2$ and $\frac{d\angle s(t)}{dt} |s(t)|^2$ are band-limited. The IF of $s(t)$ is the derivative of the phase of $s(t)$ and is simply $\omega_c + \dot{\hat{\alpha}}(t) - \dot{\hat{\beta}}(t)$, (where the dot stands for the first derivative) i.e. it consists of a d.c (corresponding to carrier frequency) and a sum of IFs of $s(t)$'s MinP and MaxP components. Clearly the behavior of the IF depends on the pole/zero locations of the signal $s(t)$ (for details see [7]). It is well-known that an analytic signal's IF could very well be negative. We now define an AllP signal whose IF is strictly positive and show that an arbitrary signal $s(t)$ observed over T seconds can be decomposed into a MinP signal and an AllP signal.

4. Periodic Signals with Positive Instantaneous Frequency

Consider a signal, $z(t)$, which is a ratio of two elementary signals as follows:

$$z(t) = \frac{1 - qe^{j\Omega t}}{1 - \frac{1}{q^*}e^{j\Omega t}};\tag{13}$$

'*' denotes complex conjugation, $q = |q|e^{j\phi}$, and $|q| > 1$. Clearly, the above expression resembles the system function of an all-pass filter [6]. Simplifying the above expression, we find that $|z(t)|$ is a constant (equal to $|q|$) for all time and hence they are called All-Phase (AllP) signals. Plugging the expressions corresponding to the identities (Eq. 6 and Eq. 7) in Eq. (13) and taking the derivative of $z(t)$'s phase angle we get

$$\frac{d\angle z(t)}{dt} = \Omega\left(1 + 2\sum_{k=1}^{\infty}|1/q_i|^k \cos(k\Omega t + k\phi)\right).\tag{14}$$

Since the right side of the expression in Eq. 14 is $\Omega(1 - |1/q|^2)|1 - 1/q^*e^{j\Omega t}|^{-2}$ and is analogous to a 'power spectrum' $z(t)$'s IF is always positive. We may generalize this result to the case of a signal consisting of a product of rational signals. If $z(t)$ is of the form

$$z(t) = \prod_{i=1}^{L}\frac{1 - q_i e^{j\Omega t}}{1 - \frac{1}{q_i^*}e^{j\Omega t}}.\tag{15}$$

then the phase angle contribution due to each of the L terms in the above equation adds up, and the corresponding IF is

$$\frac{d\angle z(t)}{dt} = \Omega\sum_{i=1}^{L}\left(1 + 2\sum_{k=1}^{\infty}|1/q_i|^k \cos(k\Omega t + k\phi_i)\right).\tag{16}$$

Since each of the L terms in the above summation is positive, the IF of the entire signal $z(t)$ is positive. These results are analogous to well known results in discrete time all-pass systems, where the equivalent of IF is the group delay; our derivation is slightly different than the one given in Oppenheim and Schafer (page 238) [6].

Now consider $s(t)$ given by Eq. 2. It can be alternatively expressed as

$$s(t) = a_0 \underbrace{\prod_{i=1}^{P}(1 - p_i e^{j\Omega t})\prod_{i=1}^{Q}(1 - \frac{1}{q_i^*}e^{j\Omega t})}_{\text{MinP}}\underbrace{\frac{\prod_{i=1}^{Q}(1 - q_i e^{j\Omega t})}{\prod_{i=1}^{Q}(1 - \frac{1}{q_i^*}e^{j\Omega t})}}_{\text{AllP}};\tag{17}$$

Again, this representation is analogous to the unique decomposition of a linear system into its minimum phase and all-pass parts. Hence, similar to $s(t)$'s representation given by Eq. 12, the signal can be expressed as a product of a MinP signal and an AllP signal as

$$s(t) = A_c e^{\alpha(t)+\beta(t)+j\left(\hat{\alpha}(t)+\hat{\beta}(t)\right)} e^{j\left(\omega_c t - 2\hat{\beta}(t)\right)}, \tag{18}$$

where $e^{j\left(\omega_c t - 2\hat{\beta}(t)\right)}$ is an AllP signal. The main point is that an analytic signal can be characterized by its positive envelope (in the traditional sense) and by a positive IF (of its AllP part) rather than by its usual IF (phase-derivative). Intuitively, this characterization tells us to define a signal's IF as the derivative of that part of its phase which is left over after removing the contribution due to the signal's log-envelope (specifically its Hilbert transform) from the original phase. Thus, given $s(t) = a(t)e^{j\phi(t)}$ we define its positive IF (PIF) as

$$\text{PIF of } s(t) = \frac{d\left(\phi(t) - \widehat{\ln a(t)}\right)}{dt}. \tag{19}$$

In [3], Loughlin and Tracer have addressed similar splitting of a signal's phase into two parts. One of their phases is obtained by integrating an FM computed as the first conditional moment of a manifestly positive joint Time-Frequency density. The other one (referred to as PM) is the phase of a residual signal obtained by coherently demodulating the given signal by the first phase; the residue's envelope is defined as AM. Although it is interesting that the concept of phase-splitting has been proposed in [3], in general their procedure is not unique: it depends on the density used for FM's computation and further the PM may not be related to AM. However, if the joint density is such that the FM computed in [3] happens to be the PIF defined by us above, then the PM will be Hilbert transform of logarithm of AM. In the following section, we propose a generalized AM-FM demodulator to compute the PIF of an analytic signal. Remarkably, the algorithm does not require explicit computation of either the logarithm or the Hilbert transform. More details will be given in [8].

5. A General AM-FM Demodulator

The proposed AM-FM demodulator, shown in Figure 1, consists of two parts. In the first part we model the envelope of the signal $s(t)$ (see Eq. (18)) by minimizing the energy of $e(t)$, defined as

$$\int_0^T |e(t)|^2 dt = \int_0^T |s(t)h(t)|^2 dt \tag{20}$$

where $h(t) = 1 + \sum_{k=1}^H h_k e^{jk\Omega t}$ is a voltage controlled oscillator (VCO) output. The minimization is achieved by choosing the coefficients, h_ks; $\Omega = 2\pi/T$. One may recognize this signal envelope modeling method as the analog of the linear prediction (autocorrelation) method well known in spectral analysis [9]. We call our method Linear Prediction in Spectral Domain or LPSD. Similar to the MinP property of the prediction error filter used in linear

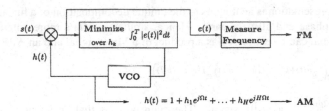

Figure 1. Block diagram for the AM-FM demodulator; AM in Figure corresponds to the traditional envelope of $1/h(t)$ (*i.e.* $1/|h(t)|$) while FM denotes the positive Instantaneous Frequency of $s(t)$'s All-Phase component.

prediction ([9]), it can be shown that minimizing $\int_0^T |e(t)|^2 dt$ will result in a $h(t)$ that is a MinP signal (having all its zeros inside the unit-circle). The significance of this property is that $h(t)$'s log-envelope and phase are Hilbert transforms. Because the error minimization is performed to approximate $s(t)$'s envelope, if the value of H is chosen sufficiently large, then $h(t)$ will be given by

$$h(t) \approx e^{-(\alpha(t)+\beta(t))} e^{-j\left(\hat{\alpha}(t)+\hat{\beta}(t)\right)}. \tag{21}$$

Thus, $\frac{1}{h(t)}$ is the desired approximation to $s_{\text{MinP}}(t)$. Consequently the error signal $e(t)$ will be

$$e(t) \approx A_c e^{j\left(\omega_c t - 2\hat{\beta}(t)\right)}, \tag{22}$$

and hence is an approximation to the AllP signal; the PIF can be obtained as $\frac{\dot{e}(t)}{|e(t)|}$ or $\frac{d\angle e(t)}{dt}$.

For example, consider a signal having 8 zeros (shown in Fig. 2(a)) and a magnitude-spectrum as shown in Fig. 2(b). The signal is sampled at 16 kHz, has 9 harmonically related complex sinusoids of frequencies 0 Hz, 200 Hz, upto 1.6 kHz, with amplitudes 1, 3.37, 3.42, 9.45, 15.76, 5.4, 3.72, and 1.5 respectively, and whose respective phases (in radians) are 0, -0.3, -1.3, -3.1, 2.8, 2.7, -1.3, -0.9, and -0.6. The signal's duration is 10 msecs (corresponding to two periods of 160 samples). We compute this signal's PIF using LPSD with model order $H = 15$.

Observe that while the signal's IF (Fig. 2(c)) is negative at times, the PIF (solid line in Fig. 2(d)) is always positive; dashed-dotted line in Fig. 2(d) corresponds to the true PIF. Although the example considered was of a periodic signal, in general the log-envelope and PIF can be estimated by applying LPSD over successive windowed portions of any given signal.

6. Conclusion

We showed that an analytic signal viewed through a window of T seconds can be uniquely decomposed into its MinP and AllP components; it can thus be represented by its (MinP's)

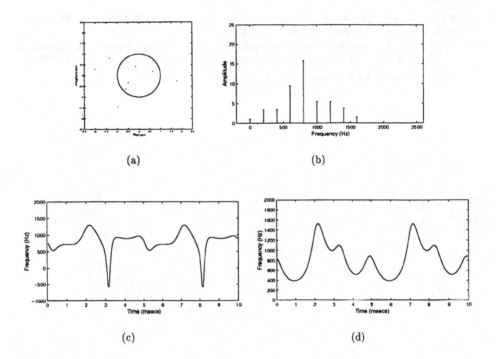

Figure 2. We consider a MixP signal consisting of 8 zeros (shown in Fig. 2(a)) having a magnitude-spectrum as shown in Fig. 2(b). The signal's IF (Fig. 2(c)) is negative around 3 msecs and 8 msecs. Its PIF displayed in Fig. 2(d) is always positive; the PIF estimated using LPSD (solid line in Fig. 2(d)) and the true PIF (dashed-dotted line in Fig. 2(d)) match closely.

positive envelope and by its (AllP's) PIF. The algorithm we proposed for this purpose does not require explicit computation of either the logarithm or the Hilbert transform.

References

1. D. Gabor, "Theory of Communication," *Proc. IEE*, vol. 93, 1946, pp. 429–457.

2. L. Cohen, *Time-Frequency Analysis*, Englewood Cliffs, NJ: Prentice Hall, 1995.

3. P. Loughlin, and B. Tacer, "On the Amplitude- and Frequency-Modulation Decomposition of Signals," *J. Acous. Soc. Amer.*, vol. 100, no. 3, 1996, pp. 1594–1601.

4. B. Boashash, "Estimating and Interpreting the Instantaneous Frequency of a Signal," *Proc. IEEE*, vol. 80, 1992, pp. 520–568.

5. H. Voelcker, "Towards a Unified Theory of Modulation Part I: Phase-Envelope Relationships," *Proc. IEEE*, vol. 54, no. 3, 1966, pp. 340–354.

6. A. Oppenheim, and R. Schafer, *Discrete-Time Signal Processing*, Englewood Cliffs, NJ: Prentice-Hall, 1989.

7. A. Rao, "Signal Analysis Using Product Expansions Inspired by the Auditory Periphery," Ph.D. Dissertation, Dept. of Elec. Engg. University of Rhode Island, Kingston, RI, 1997.

8. R. Kumaresan and A. Rao, "Model-based Approach to Envelope and Positive Instantaneous Frequency of Signals with Application to Speech," to appear in the Journal of Acoustical Society of America, 1998.

9. S. Kay, *Modern Spectral Estimation*, Englewood Cliffs, NJ: Prentice Hall, 1987.

Multidimensional Systems and Signal Processing, 9, 419–424 (1998)
© 1998 Kluwer Academic Publishers, Boston. Manufactured in The Netherlands.

Do Bounded Signals Have Bounded Amplitudes?

PATRICK J. LOUGHLIN*
Dept. of Electrical Engineering, University of Pittsburgh, Pittsburgh, PA 15261

Received November 12, 1996; Revised July 9, 1997

Abstract. There are many possible amplitudes and phases, or amplitude-modulations and frequency-modulations (AM and FM), that can generate a given signal. Of the many methods available for determining the AM and FM of a signal, some yield an unbounded AM for some magnitude-bounded signals (including the analytic signal method). The consequences of requiring a bounded AM given a bounded signal on the method used to determine the AM and FM are examined.

Key Words: amplitude modulation, frequency modulation, instantaneous amplitude, instantaneous frequency, Hilbert transform, analytic signal.

1. Introduction

Varying, or modulating, the amplitude and/or the frequency of a wave is a natural and common means of information transmission. Often, the wave is mathematically modeled as,

$$x(t) = a(t) \cos(\phi(t)) \tag{1}$$

where $a(t)$ is then said to be the amplitude modulation (AM) or "amplitude-modulation amplitude" of the wave, $\phi(t)$ is the phase, and $d\phi(t)/dt = \dot{\phi}(t)$ is the "instantaneous frequency" or frequency modulation (FM) of the wave. However, this representation is clearly not unique, for there are many pairs of functions that can play the role of amplitude and phase and which generate the identical signal. That is, one could just as easily write

$$x(t) = b(t) \cos(\theta(t)) \tag{2}$$

where $a(t) \neq b(t)$ and $\phi(t) \neq \theta(t)$. Indeed, the number of possible amplitude-phase pairs for a given signal is unlimited. Given a signal, how then do we decide on its amplitude and phase?

Gabor provided a method for specifying a particular amplitude-phase pair by associating with the given real signal $x(t)$ a complex signal

$$z(t) = A(t)e^{j\varphi(t)} = x(t) + jy(t), \tag{3}$$

where the imaginary (or quadrature) part $y(t) = H[x(t)]$ is equal to the Hilbert transform of $x(t)$ [11]. Such signals are called analytic signals, the spectrum of which is twice the spectrum of $x(t)$ for frequencies $\omega > 0$, and zero for frequencies $\omega < 0$.

---

* Supported by NSF grant MIP-9624089 and ONR grant N00014-96-1-0886.

Specifying $y(t)$ specifies the instantaneous amplitude, phase and frequency, which are given by:

$$\varphi(t) = \mathrm{atan}(y(t)/x(t)) \tag{4}$$

$$\dot{\varphi}(t) = \tfrac{d}{dt}\,\mathrm{atan}(y(t)/x(t)) \tag{5}$$

$$A(t) = (x(t) + jy(t))c^{-j\varphi(t)} = x(t)\cos(\varphi(t)) + y(t)\sin(\varphi(t)) \tag{6}$$

$$|A(t)| = \sqrt{x^2(t) + y^2(t)} \tag{7}$$

2. Why the Analytic Signal?

Gabor's method is not the only one that has been proposed for defining the instantaneous amplitude, phase and frequency of a signal. Other methods (e.g., Shekel's method [21], Mandelstam's method [25]) generally yield a different quadrature signal $y(t)$, and hence a different instantaneous amplitude, phase and frequency. What guidance is there to choose one set over all others? As Vakman notes, "one sometimes gets an evident absurdity—the instantaneous frequency takes on imaginary values, or varies from 0 to ∞ in a period" [26] and those choices can then be ruled out. But what about the remaining choices that aren't clearly absurd? There is a rich history to the problem, spanning several decades, with active debate and contributions continuing today [1–8], [10–28].

While space constraints preclude in-depth discussion of this history here, a few of the many contributions are noted. In particular, Ville [28] considered the relationship between instantaneous frequency and the Wigner distribution (later extended to other time-frequency distributions (TFDs) by Claasen and Mecklenbrauker [3]), and showed that the time-average of instantaneous frequency equals the average spectral frequency of the signal. The relationship between instantaneous frequency and spectral frequency was further explored by Mandel [16] and Fink [10], who showed that the variance of instantaneous frequency is less than the variance of spectral frequency. Bedrosian [1] and Nuttal [18] gave conditions such that $H[A(t)\cos(\varphi(t))]$ equals $A(t)\sin(\varphi(t))$, and Picinbono [19] considered the analytic signal of $A\cos(\varphi(t))$ and the conditions on the phase necessary to obtain constant instantaneous amplitude.

Vakman, among his many contributions to the subject and in particular to the study of the analytic signal, proposed reasonable physical conditions on the instantaneous amplitude, phase and frequency one obtains from the signal, and showed that only the analytic signal satisfies these conditions [25]. They are:

A. *Amplitude continuity:* a small, incremental change in the amplitude of the real signal $x(t)$ should induce a correspondingly small, incremental change in the instantaneous amplitude $A(t)$. Mathematically, this condition is expressed as

$$x(t) + \delta x(t) \Rightarrow A(t) + \delta A(t). \tag{8}$$

B. *Phase independence of scaling and homogeneity:* if the signal is scaled in amplitude by a positive, real constant, then the instantaneous phase and frequency should be unaffected, and the instantaneous amplitude should scale accordingly. Mathematically, we have

$$\text{if } x(t) \;\Rightarrow\; A(t), \varphi(t) \tag{9}$$

$$\text{then } cx(t) \;\Rightarrow\; cA(t), \varphi(t). \tag{10}$$

C. *Harmonic correspondence:* a simple sinusoid of constant frequency $\omega_0 > 0$, amplitude $A > 0$ and phase Φ should retain those values. Mathematically,

$$A\cos(\omega_0 t + \Phi) \Rightarrow A(t) = A, \quad \varphi(t) = \omega_0 t + \Phi, \quad \dot{\varphi}(t) = \omega_0. \tag{11}$$

Vakman proves that these three conditions force the quadrature signal $y(t)$ to be the Hilbert transform of the given $x(t)$, and thus only the analytic signal satisfies all three conditions for all signals; any other method or complex representation fails to satisfy at least one of these conditions for some signals [25]. Hence, the analytic signal would appear to hold the revered position as the only seemingly physically consistent method for determining the instantaneous amplitude and frequency of a signal.

3. An Alternative Consideration

Vakman's approach raises the problem of determining the instantaneous amplitude and frequency above the realm of simple mathematical definition. His physical conditions are reasonable and sensible; however, they are not the only ones that can be imposed. Consider, for example, the following amplitude condition, which is also physically reasonable, but as we show eliminates the Hilbert transform as the operator to use for defining the instantaneous amplitude, phase and frequency of a signal:

A*. *Bounded signal, bounded amplitude:* a signal bounded in magnitude should yield an instantaneous amplitude also bounded in magnitude. Mathematically,

$$|x(t)| < \infty \Rightarrow |A(t)| < \infty. \tag{12}$$

Coupled with Vakman's other two conditions (B and C), one no longer obtains the analytic signal in general, *i.e.*, while conditions A,B,C yield the Hilbert transform as the method for obtaining the instantaneous amplitude and frequency of a signal, conditions A*,B,C do not! To show this, we show that the Hilbert transform of a magnitude-bounded signal is itself not necessarily magnitude-bounded (and thus condition A* can be violated by the analytic signal). A specific example is then given.

Let $x(t)$ be a magnitude-bounded signal, i.e., $max|x(t)| = B < \infty$, with Hilbert transform $y(t)$. It follows that the magnitude of the Hilbert transform of $x(t)$ satisfies the following inequality relation:

$$|y(t)| = \left| \frac{1}{\pi} \int_{-\infty}^{\infty} \frac{x(\tau)}{t-\tau} d\tau \right| \leq \frac{1}{\pi} \int_{-\infty}^{\infty} \left| \frac{x(\tau)}{t-\tau} \right| d\tau \leq \frac{B}{\pi} \int_{-\infty}^{\infty} \left| \frac{1}{t-\tau} \right| d\tau. \tag{13}$$

But the last integral above is unbounded. Hence $|y(t)|$ is not necessarily bounded despite bounded $|x(t)|$, and therefore neither is $|A(t)|$ necessarily bounded. Indeed, the instantaneous amplitude determined per the analytic signal will generally be unbounded whenever there is a discontinuity in the bounded signal $x(t)$ [22,26].

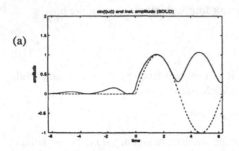
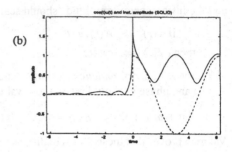

Figure 1. The magnitude of the instantaneous amplitude (solid) per the Hilbert transform of (a) $\sin(t)u(t)$ (dashed) is bounded, while that of (b) $\cos(t)u(t)$ (dashed) is unbounded at $t = 0$. Note also that the instantaneous amplitude begins before the signal does.

For example, consider the analytic signal of $\sin(\omega_0 t)u(t)$ (where $u(t)$ is the unit-step function) versus that of $\cos(\omega_0 t)u(t)$. The Hilbert transforms of these two signals are

$$H[\sin(\omega_0 t)u(t)] = \frac{1}{\pi}[\sin(\omega_0 t)\operatorname{ci}(\omega_0 t) - \cos(\omega_0 t)(\operatorname{si}(\omega_0 t) + \pi)] \tag{14}$$

$$H[\cos(\omega_0 t)u(t)] = \frac{1}{\pi}[\cos(\omega_0 t)\operatorname{ci}(\omega_0 t) + \sin(\omega_0 t)(\operatorname{si}(\omega_0 t) + \pi)] \tag{15}$$

where $\operatorname{si}(v) = -\int_v^\infty \frac{\sin \tau}{\tau} d\tau$ and $\operatorname{ci}(v) = -\int_v^\infty \frac{\cos \tau}{\tau} d\tau$ are the sine-integral and cosine-integral functions, respectively. Note that $\operatorname{si}(v)$ is bounded, while $\operatorname{ci}(v)$ is not—in particular, $|\operatorname{ci}(0)|$ is infinite.[1] It follows that $H[\cos(\omega_0 t)u(t)]$ (eq. (15)) is unbounded at $t = 0$, and hence the instantaneous amplitude (per the analytic signal) of $\cos(\omega_0 t)u(t)$ is unbounded, while that of $\sin(\omega_0 t)u(t)$ is bounded (e.g., Figure 1). Similar results hold for the more general case of $H[\cos(\omega_0 t + \phi)u(t)]$, i.e. it is unbounded except for $\phi = \pm\frac{\pi}{2}$ [22]. Note further that the envelope (i.e., instantaneous amplitude) is not causal, despite a causal signal.

4. Discussion and Conclusion

While the instantaneous amplitude and frequency of a wave are conceptually important descriptive physical characteristics of the wave, determining these quantities is a challenging problem. Indeed, it is fair to say that there is no universally agreed upon definition or method. Vakman arrived at the analytic signal as *the* method for defining the instantaneous amplitude and frequency by proposing three physical conditions for the AM and FM, and proving that only the Hilbert transform yields an AM and FM that satisfy these conditions for all signals. However, as shown, other reasonable conditions (e.g., bounded amplitude) lead to different methods.[2]

It is interesting to consider further the change on the quadrature signal $y(t)$ (or rather its functional relation to the given signal $x(t)$) that the bounded amplitude condition affects.

Recall that the quadrature signal is related to the given signal through some operation, $y(t) = H[x(t)]$ which for the moment we will not assume is the Hilbert transform (consistent with Vakman's approach). In applying conditions A,B,C and arriving at the Hilbert transform for the operator H, Vakman first showed that conditions A and B force H to be linear. He then showed that in the space of linear operators, only the Hilbert transform satisfies condition C. Consequently, in replacing A by A* and applying conditions A*,B,C, it must be that conditions A* and B no longer force a linear operator (for if they did, then condition C would again yield the Hilbert transform for H). Therefore, A*,B,C specify a nonlinear operation in general for obtaining the complex representation of the given signal.

Recently, a method was developed for obtaining an instantaneous amplitude and frequency consistent with conditions A*,B,C [14]. Per this method, the instantaneous frequency is obtained from the conditional mean frequency of a Cohen-Posch-Zaparovanny, or positive, TFD, of the signal [8,9] (which are highly nonlinear functions of the signal, and also signal-dependent distributions), and the amplitude is obtained via time-varying coherent demodulation. At present, however, there is no unique definition for the TFD (positive or otherwise) of a signal, and thus it appears that the proposed method admits more than one amplitude-phase pair for a given signal (although we note that many positive TFDs can yield the same conditional mean frequency, and hence would all yield the same amplitude-phase pair for the signal). The implication is that while conditions A,B,C specify a unique linear operator H for defining the instantaneous amplitude and frequency, conditions A*,B,C apparently specify more than one nonlinear operator.

In closing, if we accept conditions A,B,C, then we must accept that the Hilbert transform is the only way to determine the instantaneous amplitude, phase and frequency of all signals, and that sometimes these quantities may not be as physically appealing and comfortable as we expect [25]. Likewise, if we accept other physical conditions (e.g., bounded amplitude, finite support), then we must accept methods other than the Hilbert transform, which are generally nonlinear, for determining the instantaneous amplitude, phase and frequency of a signal.

Notes

1. See, for e.g., Abromowitz and Stegun, *Handbook of Mathematical Functions*, Dover Publ. Inc., NY 1965. One can also show the unboundedness of $H[\cos(\omega_0 t)u(t)]$ by computing the Hilbert transform in the spectral domain and noting that the area under the spectrum for $\omega > 0$ of $\cos(\omega_0 t)u(t)$ (which gives the value of $H[\cos(\omega_0 t)u(t)]$ at $t = 0$) is unbounded (the spectrum is $\sim 1/\omega$ for $\omega \gg \omega_0$). Conversely, the area under the spectrum for $\omega > 0$ of $\sin(\omega_0 t)u(t)$ is finite (the spectrum is $\sim 1/\omega^2$ for $\omega \gg \omega_0$).

2. Another condition that leads to a different method is that the time support of the instantaneous amplitude equal the time support of the signal ("finite time support"), which the Hilbert transform violates (e.g., Fig. 1, and also Fig. 10 in [25].)

References

1. E. Bedrosian, "A Product Theorem for Hilbert Transforms," *Proc. IEEE*, vol. 51, 1963, pp. 868–869.

2. J. Carson and T. Fry, "Variable Frequency Electric Circuit Theory with Application to the Theory of Frequency Modulation," *Bell Syst. Tech. J.*, vol. 16, 1937 pp. 513–540.

3. T. Claasen and W. Mecklenbrauker, "The Wigner Distribution—A Tool for Time-Frequency Analysis (Parts I & III)" *Phil. J. Res.*, vol. 35, 1980, pp. 217–250, pp. 372–389.

4. L. Cohen, "Distributions Concentrated Along the Instantaneous Frequency," *SPIE vol. 1348 Adv. Sig. Proc. Algs., Architect. and Implement.*, 1990, pp. 149–157.

5. L. Cohen, "Instantaneous 'Anything'," *IEEE Proc. ICASSP'93*, vol. IV, 1993, pp. 105–108.

6. L. Cohen, *Time-Frequency Analysis*, Prentice-Hall, 1995, Ch. 2.

7. L. Cohen and C. Lee, "Instantaneous Frequency, Its Standard Deviation and Multicomponent Signals," *SPIE vol. 975 Adv. Algs. and Architect. for Sig. Proc. III*, 1988, pp. 186–208.

8. L. Cohen and T. Posch, "Positive Time-Frequency Distribution Functions," *IEEE Trans. ASSP*, vol. 33, 1985, pp. 31–37.

9. L. Cohen and Y. Zaparovanny, "Positive Quantum Joint Distributions," *J. Math. Phys*, vol. 21, no. 4, 1980, pp. 794–796.

10. L. Fink, "Relations Between the Spectrum and Instantaneous Frequency of a Signal" [English translation], *Problemy Peredachi Informatsii*, vol. 2, no. 4, 1966, pp. 26–38.

11. D. Gabor, "Theory of Communication," *Jour. IEE*, vol. 93, 1946, pp. 429–457.

12. M. Gupta, "Definition of Instantaneous Frequency and Frequency Measurability," *Am. J. Phys.*, vol. 43, no. 12, 1975, pp. 1087–1088.

13. P. Loughlin and B. Tacer, "Comments on the Interpretation of Instantaneous Frequency," *IEEE Sig. Proc. Lttrs.*, vol. 4, no. 5, 1997, pp. 123–125.

14. P. Loughlin and B. Tacer, "On the Amplitude- and Frequency-Modulation Decomposition of Signals," *J. Acoust. Soc. Am.*, vol. 100, no. 3, 1996, pp. 1594–1601

15. P. Loughlin and B. Tacer, "Instantaneous Frequency and the Conditional Mean Frequency of a Signal," *Signal Processing*, vol. 60, 1997, pp. 153–162.

16. L. Mandel, "Interpretation of Instantaneous Frequencies," *Am. J. Phys.*, vol. 42, 1974, pp. 840–846.

17. P. Maragos, J. Kaiser and T. Quatieri, "Energy Separation in Signal Modulations with Application to Speech Analysis," *IEEE Trans. Sig. Proc.*, vol. 41, no. 10, 1993, pp. 3024–3051.

18. A. Nuttal, "On the Quadrature Approximation to the Hilbert Transform of Modulated Signals," *Proc. IEEE*, vol. 54, 1966, pp. 1458–1459.

19. B. Picinbono, "On Instantaneous Amplitude and Phase of Signals," *IEEE Trans. Sig. Proc.*, vol. 45, no. 3, 1997, pp. 552–560.

20. A. Rihaczek, "Hilbert Transforms and the Complex Representation of Real Signals," *Proc. IEEE*, vol. 54, 1966, pp. 434–435.

21. J. Shekel, " 'Instantaneous' Frequency," *Proc. IRE*, 1953, p. 548.

22. A. Smolinski, "On the Hilbert Envelope of a High Frequency Pulse," [English translation], *Bulletin De L'Academie Polonaise Des Sciences*, vol. XIX, no. 6, 1971.

23. D. Vakman, "On the Definition of Concepts of Amplitude, Phase and Instantaneous Frequency of a Signal," *Trans. Radio Eng. and Electron. Phy.*, 1972, pp. 754–759.

24. D. Vakman, "Do We Know What are the Instantaneous Frequency and Instantaneous Amplitude of a Signal?," *Trans. Radio Eng. and Electron. Phy.*, vol. 21, 1976, pp. 95–100.

25. D. Vakman, "On the Analytic Signal, the Teager-Kaiser Energy Algorithm, and Other Methods for Defining Amplitude and Frequency," *IEEE Trans. Sig. Proc.*, vol. 44, no. 4, 1996, pp. 791–797.

26. D. Vakman and L. Vainshtein, "Amplitude, Phase, Frequency—Fundamental Concepts of Oscillation Theory," [English translation], *Sov. Phys. Usp.*, vol. 20, no. 12, 1977.

27. B. Van der Pol, "The Fundamental Principles of Frequency Modulation," *J. IEE*, vol. 93, 1946, pp. 153–158.

28. J. Ville, "Theorie et Applications de la Notion de Signal Analytique," *Cables et Transmissions*, vol. 2A, no. 1, 1948, pp. 61–74. [English translation by RAND Corp., Tech. Rep. T-92, Santa Monica, CA, 1958.]

Multidimensional Systems and Signal Processing, 9, 425–432 (1998)
© 1998 Kluwer Academic Publishers, Boston. Manufactured in The Netherlands.

A Time-Recursive Algorithm for the Computation of Auto-Ambiguity Function and the Wigner Distribution

CHAO LU*
Department of Computer & Information Sciences, Towson State University, Baltimore, MD 21204

NING MA
Department of Computer Science & Electrical Engineering, UMBC, 1000 Hilltop Circle, Catonsville, MD 21250

JOEL M. MORRIS   morris@umbc2.umbc.edu
Department of Computer Science & Electrical Engineering, UMBC, 1000 Hilltop Circle, Catonsville, MD 21250

Received July 20, 1996; Revised February 6, 1997

Abstract. In this paper, a time-recursive algorithm for the computation of discrete-time, discrete-frequency, auto-ambiguity function is given, which in turn is used to compute the Wigner time-frequency distribution. We use the Richman-Parks-Shenoy's definition of the discrete-time, discrete-frequency, auto-ambiguity function for finite data sequences. Block diagram implementation of the time-recursive algorithm is described, computational complexity with respect to increasing block size is compared with straight forward computation, and the proper time-recursive block size is suggested.

Key Words: Time-recursive Algorithms, FFT, Gabor Transform, Time-Frequency Distribution, Wigner Distribution, Auto-Ambiguity Function.

1. Introduction

Time-recursive structures have been developed for computing the discrete Fourier transform [1,2], discrete cosine/ sine/ Hartley transforms [3], as well as the time-varying Fourier transform [4]. All of the time-recursive algorithms proposed in these papers share a common characteristic, *i.e.*, when the moving-window moves forward one step, a new sample is included while excluding the oldest sample. In [5], we proposed the block time-recursive Gabor transform. The Gabor transform has been used to characterize several versions and modifications of the Wigner distribution and related time-frequency distributions [6,7]. Fast algorithms were derived for the block time-recursive Gabor transform through two different approaches, one by the finite Zak transform, the other by biorthogonal functions. All the block time-recursive algorithms for the computation of generalized Gabor transforms can be unified by a parallel lattice structure. The unified lattice structure computes the transformed data sequentially with data sliding into/out-of a block of size N at a time.

In this paper, we apply the time-recursive idea [1,2,3] to the computation of the ambiguity function and related time-frequency distributions [8,9]. In particular, a time-recursive structure for the discrete-time, discrete-frequency, auto-ambiguity function [10,11] is presented, which in turn is used to compute the Wigner distribution. The Richman-Parks-Shenoy's

---

* This work was supported by a ONR Grant #N00014-89-J1210.

definition of the discrete-time, discrete-frequency auto-ambiguity function [11] for finite data sequences is used. Block diagram implementation of the time-recursive algorithm is described. The computation blocks are modular and independent, and the algorithm is very attractive for parallel computation. Computational complexity is discussed. In section 2, we discuss the time-recursive auto-ambiguity function and its relationship to the Wigner distribution. In section 3, block processing of the time-recursive algorithm is proposed, computational complexity with respect to increasing block size is compared with the complexity for straight forward computation, and the proper time-recursive block size is then suggested. Finally, in section 4, the main contribution of this paper is summarized. These techniques are applicable to the fast computation of a variety of time-frequency distributions [12,13,14].

2. Time-Recursive Auto-Ambiguity Function and Wigner Distribution

For a finite data sequence $\underline{x} = \{x(0), x(1), x(2), \ldots, x(L-1)\}$, the discrete-time, discrete-frequency, auto-ambiguity function [11] is

$$A_x(m, n) = \sum_{l=0}^{L-1} e^{\frac{j\pi(mn)_L}{L}} e^{\frac{j2\pi ml}{L}} x(l)x^*((l+n)_L), \quad 0 \le m, n < L \tag{1}$$

in which the notation $(a)_L$ means $a \bmod L$. Then the auto-Wigner distribution of $\underline{x}$ will be the two-dimensional Fourier transform of the auto-ambiguity function,

$$
\begin{aligned}
W_x(\alpha, \beta) &= F_{2D}(L, L)A_x \\
&= \frac{1}{L} \sum_{m=0}^{L-1} \sum_{n=0}^{L-1} \sum_{l=0}^{L-1} e^{\frac{-j2\pi(\alpha m + \beta n)}{L}} e^{\frac{j\pi(mn)_L}{L}} e^{\frac{j2\pi ml}{L}} x(l)x^*((l+n)_L), \quad 0 \le \alpha, \beta < L.
\end{aligned} \tag{2}
$$

We can multiply both sides of Eq. (1) by $e^{\frac{-j\pi(mn)_L}{L}}$, and denote the result as

$$B_x(m, n) = e^{\frac{-j\pi(mn)_L}{L}} A_x(m, n) = \sum_{l=0}^{L-1} e^{\frac{j2\pi ml}{L}} x(l)x^*((l+n)_L). \tag{3}$$

To compute $B_x(m, n), 0 \le m, n < L$, we first multiply the data sequence by itself $x(l)x^*(l + n)_L$, and then calculate a set of (L) L-point FFTs. Rewriting $B_x(m, n)$ in terms of time t for a shifted data sequence $\underline{x}_t = \{x(t), x(t+1), x(t+2), \cdots, x(t+L-1)\}$ in detail, we can define the auto-ambiguity function at time t as

$$B_x^t(m, 0) = \sum_{l=0}^{L-1} e^{\frac{j2\pi ml}{L}} x(l+t)x^*(l+t)$$

$$B_x^t(m, 1) = \sum_{l=0}^{L-1} e^{\frac{j2\pi ml}{L}} x(l+t)x^*((l+1)_L + t)$$

$$\vdots$$

$$B_x^t(m, L-1) = \sum_{l=0}^{L-1} e^{\frac{j2\pi ml}{L}} x(l+t)x^*((l+L-1)_L+t),$$

$$0 \le m < L. \tag{4}$$

To derive a time-recursive relationship between the updated auto-ambiguity function of the next input data sequence $\underline{x}_{t+1} = \{x(t+1), x(t+2), \cdots, x(t+L-1), x(t+L)\}$ and the previous computed one, we write $B_x(m, n)$ at time $t+1$,

$$B_x^{t+1}(m, 0) = \sum_{l=0}^{L-1} e^{\frac{j2\pi ml}{L}} x(l+t+1)x^*(l+t+1)$$

$$B_x^{t+1}(m, 1) = \sum_{l=0}^{L-1} e^{\frac{j2\pi ml}{L}} x(l+t+1)x^*((l+1)_L+t+1)$$

$$\vdots$$

$$B_x^{t+1}(m, L-1) = \sum_{l=0}^{L-1} e^{\frac{j2\pi ml}{L}} x(l+t+1)x^*((l+L-1)_L+t+1), \tag{5}$$

which can be written as

$$B_x^{t+1}(m, 0) = e^{\frac{-j2\pi m}{L}} \sum_{l=1}^{L} e^{\frac{j2\pi ml}{L}} x(l+t)x^*(l+t)$$

$$B_x^{t+1}(m, 1) = e^{\frac{-j2\pi m}{L}} \sum_{l=1}^{L} e^{\frac{j2\pi ml}{L}} x(l+t)x^*((l)_L+t+1)$$

$$\vdots$$

$$B_x^{t+1}(m, L-1) = e^{\frac{-j2\pi m}{L}} \sum_{l=1}^{L} e^{\frac{j2\pi ml}{L}} x(l+t)x^*((l+L-2)_L+t+1). \tag{6}$$

Substituting Eq. (4) into (6), we can then obtain the time-recursive auto-ambiguity function as

$$B_x^{t+1}(m, n) = e^{\frac{-j2\pi m}{L}} [B_x^t(m, n) + x(t+L-n)(x^*(t+L) - x^*(t))e^{\frac{j2\pi m(L-n)}{L}}$$
$$+ x^*(t+n)(x(t+L) - x(t))], \qquad 0 \le m, n < L. \tag{7}$$

From Eq. (7) it is clear that the auto-ambiguity function of the new data sequence $\underline{x}_{t+1}$, when a new data sample shifts in and an old data sample shifts out, is given by the auto-ambiguity function of $\underline{x}_t$ with the addition of two error-correction terms. Equation (7) is called the time-recursive algorithm (RTA) for the computation of the auto-ambiguity function.

The computational flow-chart of the time-recursive auto-ambiguity function is shown in Figure 1 (Note, $W_L(m) = e^{\frac{j2\pi m}{L}}$). From Eq. (7), we note that to calculate $B_x^{t+1}(m, n)$, for all $0 \le m, n < L$, there are $(1+2L)L$ additions and $4L^2$ multiplications. For all $A_x^{t+1}(m, n)$, L^2 additional multiplications are needed. The computational complexity of an algorithm can be characterized by the total number of additions and multiplications needed. Denote

COM_TRA as the computational complexity of the TRA, then

$$COM\_TRA = (1 + 2L)L + 4L^2 + L^2 = 7L^2 + L. \tag{8}$$

Since Eq. (4) is a set of L-point FFTs, to compute $B_x^{t+1}(m, n)$ by definition, $L(L \ log_2L)$ additions and $L(L \ log_2L + L)$ multiplications are needed. So the corresponding computational complexity COM_DIR for $A_x^{t+1}(m, n)$ can be given as

$$COM\_DIR = L(L \ log_2L) + L(L \ log_2L + L) + L^2 = 2L^2(log_2L + 1). \tag{9}$$

Using Eq. (7) we can derive a time-recursive algorithm for the computation of the Wigner distribution as

$$
\begin{aligned}
W_x^{t+1}(\alpha, \beta) &= F_{2D}(L, L)A_x^{t+1} \\
&= F_{2D}(L, L)B_x^{t+1}e^{\frac{j\pi(mn)_L}{L}} \\
&= W_x^t(\alpha + 1, \beta) + \hat{w}_x^t(\alpha, \beta),
\end{aligned}
\tag{10}
$$

where the error term $\hat{w}_x^t$ comes from the error term in B_x^{t+1} of Eq. (7), and it is given as

$$
\begin{aligned}
\hat{w}_x^t(\alpha, \beta) = {} & \frac{1}{L}[x(t + L) - x(t)]^* \sum_{n=0}^{L-1} x(t + L - n)e^{\frac{-j2\pi\beta n}{L}} \sum_{m=0}^{L-1} e^{\frac{j\pi(mn)_L}{L}} e^{\frac{-j2\pi(\alpha+n+1)m}{L}} \\
& + \frac{1}{L}[x(t + L) - x(t)] \sum_{n=0}^{L-1} x^*(t + n)e^{\frac{-j2\pi\beta n}{L}} \sum_{m=0}^{L-1} e^{\frac{j\pi(mn)_L}{L}} e^{\frac{-j2\pi(\alpha+1)m}{L}}.
\end{aligned}
\tag{11}
$$

Denote

$$f(\alpha, n) = \frac{1}{L} \sum_{m=0}^{L-1} e^{\frac{j\pi(mn)_L}{L}} e^{\frac{-j2\pi\alpha m}{L}}, \tag{12}$$

then $[f(\alpha, n)]_{0\le\alpha<L, 0\le n<L}$ is a sparse matrix with symmetries, and it can be pre-computed. Then we have

$$
\begin{aligned}
\hat{w}_x^t(\alpha, \beta) = {} & [x(t + L) - x(t)]^* \sum_{n=0}^{L-1} x(t + L - n)f(\alpha + n + 1, n)e^{\frac{-j2\pi\beta n}{L}} \\
& + [x(t + L) - x(t)] \sum_{n=0}^{L-1} x^*(t + n)f(\alpha + 1, n)e^{\frac{-j2\pi\beta n}{L}}.
\end{aligned}
\tag{13}
$$

The computational complexity of $\hat{w}_x^t(\alpha, \beta)$ is characterized by a set of 1-dimensional FFTs.

3.   Block Time-Recursive Algorithm

In the previous section, the time-recursive algorithm is based on block-size-one updating. In application, we might be interested in block processing, that is we would like to advance

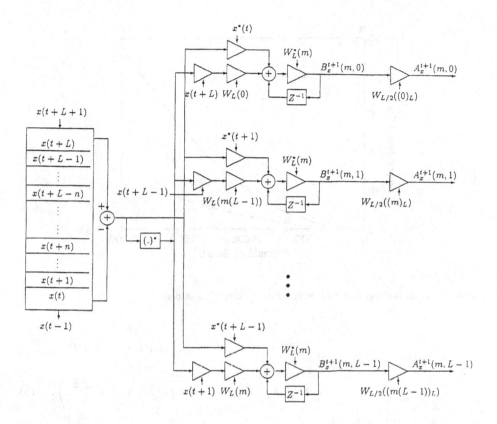

Figure 1. Implementation block diagram of the TRA.

the data sequence by a block size larger than one. This motivates the block time-recursive algorithm (BTRA) for the computation of the auto-ambiguity function or the Wigner distribution.

We begin with block-size-two updating. From Eq. (7), the block-size-two auto-ambiguity function $B_x^{t+2}(m, n)$ can be written as

$$B_x^{t+2}(m, n) = e^{\frac{-j2\pi m}{L}}[B_x^{t+1}(m, n)$$
$$+ x(t+1+L-n)(x^*(t+1+L) - x^*(t+1))e^{\frac{j2\pi m(L-n)}{L}}$$
$$+ x^*(t+1+n)(x(t+1+L) - x(t+1))], \tag{14}$$
$$0 \le m, n < L.$$

Substituting Eq. (7) into Eq. (14), then

$$B_x^{t+2}(m, n) = e^{\frac{-j2\pi m(2)}{L}}\left(B_x^t(m, n)\right.$$

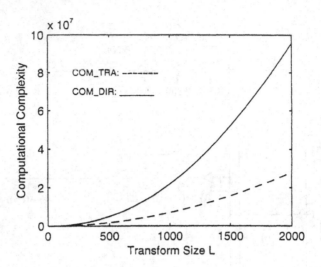

Figure 2. Complexity Comparison between the TRA and Direct Computation

$$+ \sum_{k=0}^{1} [e^{\frac{-j2\pi mn}{L}} x(t+k+L-n)(x^*(t+k+L) - x^*(t+k))$$

$$+ x(t+k+n)(x(t+k+L) - x(t+k))]e^{\frac{-j2\pi mk}{L}} \Bigg), \quad (15)$$

$0 \le m, n < L.$

Similarly, we can get the formula for the block-size-i, time-recursive, auto-ambiguity function $B_x^{t+i}(m,n)$ as

$$B_x^{t+i}(m,n) = e^{\frac{-j2\pi mi}{L}} \Bigg(B_x^t(m,n)$$

$$+ \sum_{k=0}^{i-1} [e^{\frac{-j2\pi mn}{L}} x(t+k+L-n)(x^*(t+k+L) - x^*(t+k))$$

$$+ x(t+k+n)(x(t+k+L) - x(t+k))]e^{\frac{-j2\pi mk}{L}} \Bigg), \quad (16)$$

$0 \le m, n < L.$

Recognize that the second term of Eq. (16) is a set of pruning FFTs with an input data block size of i and a Fourier transform size of L. More advanced pruning FFT algorithms [15] can be used for its computation. If we simply calculate Eq. (16) by multiply and add, then $B_x^{t+i}(m,n)$ requires $((1+L)i + (i-1)L + L)L$ additions and $(4i+1)L^2$ multiplications.

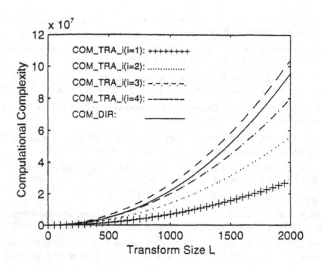

Figure 3. Comparison between *COM_TRA_i* and *COM_DIR*.

Therefore, the computational complexity COM_TRA_I of $A_x^{t+i}(m, n)$ is

$$COM\_TRA\_i = ((1 + L)i + (i - 1)L + L)L + (4i + 1)L^2 + L^2 = (6i + 2)L^2 + iL. \quad (17)$$

The comparison between COM_TRA_i and COM_DIR is shown in Figure 3.

4. Conclusions

A new scheme employing the time-recursive approach to compute the auto-ambiguity function and the Wigner time-frequency distribution is presented in this paper. The computational complexity is a linear function of the transform size L^2 in comparison with $L^2 log L$ for the straight forward computation. The trade-off between advancing block size and computational cost is discussed: block processing sizes smaller than 4 are suggested. If a fast pruning FFT algorithm is used, relatively larger blocks (larger than 4) can be advanced in the time-recursive computation. A block diagram implementation of the algorithm is presented; the structure of the algorithm is modular and parallel, and is very attractive for real-time computation via parallel implementation.

References

1. R. R. Bitmead, "On Recursive Discrete Fourier Transform," *IEEE Trans. on ASSP*, vol. 30, no. 2, 1982, pp. 319–322.

2. M. Unser, "Recursion in Short-time Signal Analysis," *Signal Processing*, vol. 5, no. 3, 1983, pp. 229–240.

3. K. J. R. Liu, and C. T. Chiu, "Unified Parallel Lattice Structures for Time-Recursive Discrete Co-sine/Sine/Hartley Transforms," *IEEE Trans. SP*, vol. 41, no. 3, 1993, pp. 1357–1377.

4. W. Chen, N. Kehtarnavaz, and T. W. Spencer, "An Efficient Recursive Algorithm for Time-Varying Fourier Transform," *IEEE Trans. SP*, vol. 41, no. 7, 1993, pp. 2488–2490.

5. C. Lu, S. Joshi, and J. M. Morris, "Block Time-Recursive Generalized Gabor Transform and Its Implementation," *Proc. Conf. Inform. Sci. & Sys.*, Princeton Univ., 20–22 March 1996, pp. 1200–1201.

6. S. Qian, and J. M. Morris, "Wigner Distribution Decomposition and Cross-term Deleted Representation," *Signal Processing*, vol. 27, no. 2, 1992, pp. 125–144.

7. J. M. Morris, and D. S. Wu, "Some Results on Joint Time-Frequency Representations via Gabor Expansions," *Proc. Conf. Inform. Sci. & Sys.*, Princeton Univ., 18–20 March 1992, pp. 6–10.

8. D. Wu, and J. M. Morris, "Discrete Cohen's Class of Distributions," *Proc. IEEE-SP Int'l. Symp. on TFTS Analysis*, Phila., PA, USA, 25–28 Oct. 1994, pp. 532–535.

9. L. Cohen, *Time Frequency Analysis.* Prentice-Hall, 1995.

10. F. Peyrin and R. Prost, "A Unified Definition for the Discrete-Time, Discrete-Frequency, and Discrete-Time/Frequency Wigner Distributions," *IEEE Trans. ASSP*, vol. ASSP-34, no. 4, 1986, pp. 858–866.

11. M. S. Richman, T. W. Parks, and R. G. Shenoy, "Discrete-Time, Discrete-Frequency Time-Frequency Representations," *Proc. IEEE ICASSP'95*, vol. 2, Detroit, MI, 9–12 May 1995, pp. 1029–1032.

12. J. Jeong, and W. J. Williams, "Alias-free Generalized Discrete-Time Time-Frequency Distributions," *IEEE Trans. SP*, vol. 40, No.11, 1992, pp. 2757–2765.

13. J. M. Morris, and D. S. Wu, "On Alias-Free Formulations of Discrete-Time Cohen's Class of Distributions," *IEEE Trans. SP*, vol. 44, no. 6, June 1996, pp. 1355–1364.

14. R. N. Czerwinski, and D. L. Jones, "Adaptive Cone Kernel Time-Frequency Analysis," *IEEE Trans. SP*, vol. 43, no. 7, July 1995, pp. 1715–1718.

15. S. He, and M. Torkelson, "Computing Partial DFT for Comb Spectrum Evaluation," *IEEE Signal Processing Letters*, vol. 3, no. 6, 1996, pp. 173–175.

Multidimensional Systems and Signal Processing, 9, 433–438 (1998)
© 1998 Kluwer Academic Publishers, Boston. Manufactured in The Netherlands.

The "Harmonic" Rejecting Correlation Function

DOUGLAS NELSON   djnelso@afterlife.ncsc.mil
Dept. of Defense, 9800 Savage Rd., Ft. Meade, Md. 20755

Received September 2, 1996; Revised March 16, 1997

Abstract. A classical problem in signal processing is accurate estimation of fundamental frequency/periodicity of periodic signals at low SNR. Typically, researchers address the estimation problem, assuming that the signal environment is a sum of sinusoids in white Gaussian noise. If the signals and noise are pulsed, the situation is much more complex since normal FFT based methods result in spectra which are sums of harmonic structures. Sorting radar signals can be especially difficult since there may be many pulsed signals present in a low SNR impulsive noise environment. In this paper, a method equivalent to integration along a hyperbola on the Wigner distribution is presented. This transform, which is closely related to both the Fourier transform and the correlation function, has the property that a periodic signal produces an expected non-zero complex-valued bulge at only the fundamental. The phase, magnitude and position of the correlation bulge are sufficient to characterize the time-domain pulse train. Finally, a simple super-resolution method is presented which may be used to refine the fundamental frequency/period estimate.

Key Words: Correlation, spectrum, Wigner distribution, frequency, period, pulse train.

Introduction

One of the most common problems in signal processing is accurate estimation of fundamental frequency/periodicity of signals at low SNR. In estimating these signal parameters, most published research is based on the assumption that the signal environment consists of a single sinusoid or a sum of sinusoids in additive white Gaussian noise. Under this assumption, the problem is greatly simplified because the expected Fourier spectrum is the superposition of one or more impulses, one impulse for each sinusoid in the environment. If the signals and noise are pulsed, as is the case in many radar environments, the situation is much more complex. This problem occurs in radar and also in speech processing, where voiced speech consists of a series of pulses which resonate in the vocal tract, making it difficult to recover the pitch excitation function [1]. For pulse trains, normal FFT based methods result in spectra which are the superposition of harmonic structures (see, for instance, [5]). In general, any periodic, non-sinusoidal signal has an expected spectrum which is a harmonic structure consisting of a sequence of bulges at integer multiples of the fundamental frequency.

The problem of signal estimation is manageable if the environment contains only one or two pulse trains; however, the problem becomes very difficult if there are many signals present and the noise is impulsive. The normal methods for processing this environment are based on the autocorrelation function

$$R(F, G, I) = \frac{1}{\|I\|} \int_I F(t)\overline{F(t + \tau)}dt, \tag{1}$$

where $\|I\|$ is the length of the interval I. The expected autocorrelation distribution for any periodic signal has the same periodicity as the original signal. This unfortunately means

101

that signal sorting is no easier in the correlation domain that it is in the frequency time domains.

We introduce the notion of subharmonics and time-harmonics. For positive integers n, signal period p and signal frequency f, the nth subharmonic is f/n, and the nth time-harmonic is np. With this notation, and the observation that periodicity and frequency are reciprocals of each other, we see that the Fourier spectrum of a periodic, non-sinusoidal signal contains harmonics but does not contain time-harmonics (sub-harmonics.) Similarly, the correlation function contains time-harmonics but not harmonics.

In radar processing, the fundamental period is normally called the Pulse Repetition Interval (PRI), which we define as the minimum p satisfying

$$X(t) = X(t + p). \tag{2}$$

For convenience, we assume a signal which is a unit impulse train of the form

$$U(t, \Phi_p, p) = \begin{cases} 1, & t = \left(n + \frac{\Phi_p}{2\pi}\right)p \\ 0 & \text{otherwise} \end{cases}, n = 1, 2, \ldots \tag{3}$$

The quantity $-\pi \le \Phi_p < \pi$ is called the PRI-phase. For unit impulse trains, the quantities p and Φ_p are uniquely determine the signal.

With notation above, we will construct a transform closely related to both the correlation function and the Fourier transform. This transform has the desirable properties of both the Fourier transform and correlation function. It produces a bulge at the fundamental period and has an expected value of zero at the harmonics and time-harmonics. The transform is complex valued, having an argument equal to Φ_p for input signals which are pulse trains. In addition, there is a simple interpolation formula based on Φ_p which may be used to accurately resolve the true period of a periodic pulse train. This interpolation provides a capability to resolve periodicity of quantized signals to much higher resolution than the time-quantization error of the signal. The method is related to the Wigner distribution in the sense that it is equivalent to integration along a hyperbola on the Wigner distribution surface.

The method developed in this paper is based on the cross-spectral methods developed in [3,4,7], which have been used to accurately recover multiple weak sinusoids in noise. The correlation method presented here performs well in estimating and resolving multiple periodic signals, not just sinusoids. The method is especially well suited to the problem of sorting superimposed pulse trains. Both this correlation method, and the cross-spectral methods are generalizations of the phase-based methods of Tretter [6], Kay [2], and others.

The Harmonic Rejecting Transform

We define the Harmonic Rejecting CrossCorrelation Function (HRCCF) of two signals F and G over interval I as

$$C_h(F, G, \tau, I) = \frac{1}{\|I\|} \int_I F(t)\overline{G(t + \tau)}e^{2\pi it/\tau}dt, \quad (\tau \ne 0). \tag{4}$$

where $\|I\|$ is the length of the interval of integration I. If $F = G$ the correlation function (4) becomes an autocorrelation function (HRACF)

$$R_h(F, \tau, I) = \frac{1}{\|I\|} \int_I F(t)\overline{F(t + \tau)}e^{2\pi it/\tau}dt, \quad (\tau \neq 0). \tag{5}$$

Correlation representations of (4) and (5) were chosen because of radar applications. We could have equally well have written the HRCCF as a frequency domain representation

$$\mathcal{F}_h(F, G, \omega, I) = \frac{1}{\|I\|} \int_I F(t)\overline{G\left(t + \frac{2\pi}{\omega}\right)}e^{i\omega t}dt, \quad (\omega \neq 0). \tag{6}$$

Equation (5) can be made symmetric by a change of variables to obtain the formula

$$R_h(F, \tau, I) = \frac{-1}{\|I'\|} \int_{I'} F\left(t - \frac{\tau}{2}\right) \overline{F\left(t + \frac{\tau}{2}\right)}e^{2\pi i\frac{t}{\tau}}dt, \quad (\tau \neq 0), \tag{7}$$

which is similar to the standard representation of the Wigner distribution

$$W(F, t, \omega, I) = \frac{1}{\|I\|} \int_I F\left(t - \frac{\tau}{2}\right) \overline{F\left(t + \frac{\tau}{2}\right)}e^{-i\omega t}d\tau. \tag{8}$$

Formula (7) is easily seen to be an integral along hyperbola on the Wigner distribution surfaces.

To derive the properties of the correlation functions, we assume that F and G are periodic with fundamental period p. We note first that the function (4) vanishes in the limit for all values of τ equal to non-unity integer multiples of p. To be precise,

$$\lim_{\|I\| \to \infty} C_h(F, G, np, I) = 0, \quad (n = 2, 3, 4, \ldots), \tag{9}$$

The proof of (9) follows immediately from the periodicity of F and G. We assume that $I = [a, b]$, $\tau = np$ and $n > 1$. We first note that, for integers k and n,

$$C_h(F, G, np, [a, a + knp]) = \frac{1}{knp}\left[\sum_{m=0}^{n-1} C_h(F, G, np, [a + mp, a + (m + 1)p])\right]$$

$$= \frac{1}{kn}C_h(F, G, np, [a, a, +p]) \sum_{m=0}^{n-1} e^{2\pi im/n}$$

$$\equiv 0. \tag{10}$$

Equation (9) then reduces to

$$C_h(F, G, np, [a, b]) = \frac{1}{\|[a, b]\|} \int_{b - \|[a,b]\|_{np}}^{b} F(t)\overline{G(t + np)}e^{2\pi it/np}dt, \tag{11}$$

where $|_{np}$ represents the remainder after division by np. Clearly, (11) converges to zero as $\|[a, b]\| \to \infty$.

The frequency harmonic annihilation property, we will only prove in the case of a periodic unit impulse trains for the autocorrelation function. This is a trivial, but important case in radar signal processing. Under this assumption, we note that

$$U(t, \Phi, p)U\left(t + \frac{p}{n}, \Phi, p\right) = 0, n = 2, 3, \ldots, \tag{12}$$

and the result follows.

For the values of $\tau \neq np$, the correlation function (4) of arbitrary periodic signals does not necessarily vanish since the correlation function can have strong partial correlation bulges at lags other than the signal period. This situation exists for both the Fourier spectrum and the normal correlation function as well as the HRCCF.

In the above discussion, the property which makes the entire process work is the introduction of the exponential factor into the correlation function. This factor essentially winds the correlation function about the unit circle, resulting in a correlation function which is the center of mass of a complex distribution. Symmetry properties of the resulting distribution guarantee that the function has a zero-valued complex mean at multiples of the signal period.

PRI-Phase and the Interpolation Formula

The PRI phase of a pulse train, with period p, is given by

$$\Phi_p = \left. \frac{2\pi T_n}{p} \right|_{2\pi} \tag{13}$$

where T_n is the Time-Of-Arrival (TOA) of the nth pulse, and $|_{2\pi}$ assumes values in the interval $[-\pi, \pi)$. If we apply the autocorrelation function (5) to a unit impulse train (3), we get

$$R_h(U, p, I) = \frac{1}{\|I\|} \int_I |U(t, \Phi_p, p)|^2 e^{2\pi it/p} dt. \tag{14}$$

Since U is a p-periodic impulse train,

$$\arg\{R_h(U, p, I)\} = \Phi_p. \tag{15}$$

showing that the argument of the HRACF applied to pulse trains is the PRI-phase of the pulse train.

Finally, we derive the interpolation formula. For this, we assume a periodic pulse train (3) and assume that the period has been improperly estimated as $\tilde{p}$, where the error in the period measurement is smaller than the observed width of the correlation bulge. We tile the time axis with non-overlapping intervals of length L

$$I_n = [nL + \alpha, (n + 1)L + \alpha] \tag{16}$$

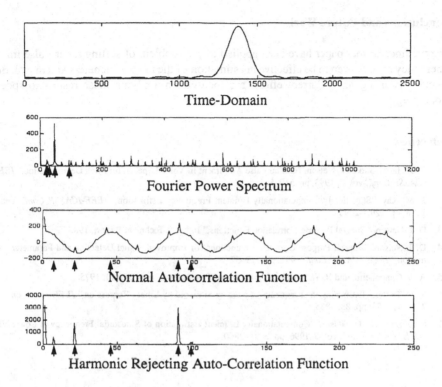

Figure 1. Sum of the five signals

$$0.1 \sin\left(\frac{2\pi t}{99.111}\right), 0.5 \sin\left(\frac{2\pi t}{45.3}\right), U(t, 0, 21), 0.1U(t, 0, 63), 3U(t, 0, 91)$$

displayed in time, frequency, autocorrelation and HRACF domains.

and calculate the expected advance in the PRI-phase error as

$$E(\Delta\Phi_p(\tilde{p})) = \arg\left\{\sum_n R_h(U(t, \Phi, p), \tilde{p}, I_{n+1})\overline{R_h(U(t, \Phi, p), \tilde{p}, I_n)}\right\}. \qquad (17)$$

Since (18) represents the average advance of the estimated signal phase over time intervals of length L, we can re-estimate the signal period as

$$p_{rest} = \tilde{p}\left(1 + \frac{E(\Delta\Phi_p(\tilde{p}))}{2\pi}\right). \qquad (18)$$

We may iterate the process to improve the accuracy of the signal period and PRI-phase.

Conclusions and Future Work

The methods of this paper have been applied to the problem of sorting radar pulse trains, where they have proven to be effective in estimation of the radar parameters and in isolating specific radar signals. Current efforts are focused on applying these results to speech processing.

References

1. J. D. Harris, and D. J. Nelson, "Glottal Pulse Alignment in Voiced Speech for Pitch Determination," *IEEE-ICASSP Conf.*, vol. 2, 1993, pp. 519–522.

2. S. M. Kay, "Statistically/Computationally Efficient Frequency Estimation," *IEEE-ICASSP Conf.*, vol. 5, 1988, pp. 2292–2295.

3. D. J. Nelson, "Special Purpose Correlation Functions," Internal Technical Report, 1988.

4. D. J. Nelson, "Special Purpose Correlation Functions for Improved Signal Detection and Parameter Estimation," *IEEE-ICASSP Conf.*, 1993, pp. 519–522.

5. A. V. Oppenheim, and R. W. Schafer, *Digital Signal Processing*, Prentice Hall, 1975.

6. S. A. Tretter, "Estimating the Frequency of a Noisy Sinusoid by Linear Regression," *IEEE Trans. on Info. Theory*, 1985, pp. 832–835.

7. S. Umesh, and D. Nelson, "Computationally Efficient Estimation of Sinusoidal Frequency at Low SNR," *IEEE-ICASSP Conf.*, vol. 5, 1996, pp. 2797–2800.

Multidimensional Systems and Signal Processing, 9, 439–445 (1998)
© 1998 Kluwer Academic Publishers, Boston. Manufactured in The Netherlands.

Positive Time-Frequency Distributions via Quadratic Programming

JAMES W. PITTON
MathSoft, 1700 Westlake Ave N, Seattle, WA 98109

Received January 15, 1997; Revised January 6, 1998

Abstract. A new method for computing positive time-frequency distributions (TFDs) for nonstationary signals is presented. This work extends the earlier work of the author and his colleagues in computing positive TFDs [8,11]. This paper describes a general quadratic programming approach to the problem of computing these signal-dependent distributions. The method is based on an evolutionary spectrum formulation of positive TFDs. The minimization problem reduces to a linearly-constrained quadratic programming problem, for which standard solutions are widely available.

Key Words: time-frequency analysis, time-varying spectral estimation, nonstationary signals, optimization, quadratic programming, nonlinear programming

I. Introduction

The spectrogram represents the traditional approach to time-frequency signal analysis. This representation is based on short-time Fourier analysis, using a sliding window to compute a time-varying spectrum. The resulting representation, however, suffers from a trade-off between time and frequency resolution, providing one at the other's expense. An improvement over traditional short-time spectral analysis is embodied in the class of bilinear time-frequency distributions [2], which seeks to accurately represent the energy of a signal jointly in time and frequency. Such an energy distribution is only approximated by the spectrogram. Commonly known as Cohen's class of bilinear time-frequency representations, the properties of a particular representation are determined by an associated kernel function. By an appropriate choice or design of the kernel, these representations are able to simultaneously satisfy the time and frequency marginals of the signal exactly, thus avoiding the time-frequency resolution trade-off of the spectrogram. One such representation is the Wigner distribution, which suffers no time-frequency trade-off and provides sharp resolution of time-varying frequencies. However, the resulting time-frequency "spectrum" is not strictly non-negative, and many aspects of the resulting time-frequency display are difficult to interpret in light of known or observed frequency content of the signal. Much of the work in time-frequency has been to find a representation with better resolution than the spectrogram, but without the drawbacks of the Wigner distribution. Many other time-frequency representations have been developed which provide improved joint time-frequency resolution over that of the spectrogram and alleviate to varying degrees the shortcomings of the Wigner distribution. However, they also generate negative values which complicate the estimation and interpretation of subsequent parameters, making interpretation as an energy spectrum difficult. These difficulties may be overcome by computing positive time-frequency distributions.

107

II. Positive Time-Frequency Distributions

A positive time-frequency distribution, or TFD, is simply a time-varying spectrum. It provides a measure of the instantaneous energy of a signal at a particular time and frequency. Unlike the Wigner distribution and related methods, it is everywhere nonnegative, providing a meaningful estimate of the signal energy. Unlike the spectrogram, it yields the correct univariate marginal distributions in time and frequency, namely the instantaneous energy and the energy spectral density. Thus, a valid TFD must at the very least satisfy these three fundamental properties of distributions:

$$P(t, \omega) \geq 0, \tag{1}$$

$$\int P(t, \omega)d\omega = |s(t)|^2, \tag{2}$$

$$\int P(t, \omega)dt = |S(\omega)|^2, \tag{3}$$

where $S(\omega)$ denotes the Fourier transform of the signal $s(t)$, and all integrals are from $-\infty$ to ∞. TFDs satisfying these three properties are also known as Cohen-Posch TFDs, after the researchers who first showed that such distributions exist for all signals [3].

The first method for generating positive TFDs was presented in [7]. Detailed in [8], this method uses constrained optimization, finding the distribution that minimizes the cross-entropy to some prior distribution, subject to a set of linear constraints. Positive TFDs were computed using spectrograms and combinations of spectrograms as prior distributions. Positivity and the marginals were used as constraints in the optimization procedure. These constraints modified the marginal properties of the distribution; joint time-frequency structure came from information contained in the prior distribution. A drawback of this approach is that flaws of the prior, e.g., the blurring of the spectrogram, can propagate through to the posterior distribution. This drawback can be alleviated somewhat by choosing an accurate prior distribution, or avoided altogether by using a uniform prior and specifying additional constraints which provide additional time-frequency information. This choice of prior is equivalent to maximum entropy estimation. Such a method was presented in [11,12]. Based on the theory of the evolutionary spectrum [13], a convolutional relationship between a spectrogram and a positive TFD was derived. Using this derived relationship as a constraint, maximum entropy estimation was used to compute the TFD. This method used spectrograms of various lengths and frequency modulation rates in the procedure, improving the resolution of time-varying signal components. Maximum entropy estimation was also used in [5] to generate positive TFDs based on frequency-modulated marginal constraints.

Least-squares estimation has also been used to compute positive TFDs. Working with the kernel formulation of a positive TFD [3], a positive TFD was obtained that minimized the L_2 norm of the distance to a pre-specified "template" subject to positivity and the marginals [16,6]. The authors used a Wigner distribution as the template, but recommended that other bilinear distributions or combinations of distributions be used. A more efficient procedure for generating positive TFDs using least-squares minimization was presented in [14]. The authors used an alternating projections procedure to minimize the squared distance to a Reduced Interference Distribution subject to positivity and the marginals.

In the next section, a general method is derived for computing positive TFDs using least-squares optimization. The previous methods of computing TFDs via least-squares offered no justification of the procedure, or guidance as to the choice of the target representation. The approach presented next directly addresses this issue. As in [11] for maximum entropy TFDs, the method is based on an evolutionary spectrum formulation of a positive TFD. Approximate constraints on the TFD are derived, and the TFD can then be computed by least-squares minimization of the error in the constraints. The minimization problem reduces to a linearly-constrained quadratic programming problem, for which standard solutions are readily available.

III. Positive TFDs via Quadratic Programming

A. Derivation of Minimization Problem

As discussed in [11], positive TFDs may be based on Priestley's formulation of the Evolutionary Spectrum [13]. Following this theory, a signal may be written as:

$$s(t) = \int A(t, \omega)e^{j\omega t}dZ(\omega) \tag{4}$$

where $dZ(\omega)$ is an orthogonal process with unit variance. $A(t, \omega)$ satisfies a statistical time marginal:

$$\int |A(t, \omega)|^2 d\omega = E\left[|s(t)|^2\right], \tag{5}$$

and may be chosen to satisfy the frequency marginal:

$$\int |A(t, \omega)|^2 dt = E\left[|S(\omega)|^2\right]. \tag{6}$$

$|A(t, \omega)|^2$ is then a valid positive time-frequency distribution:

$$P(t, \omega) = |A(t, \omega)|^2. \tag{7}$$

An estimation procedure for positive TFDs is now readily obtained. The ambiguity function $\chi(\theta, \tau)$ of a signal $s(t)$ is defined as

$$\chi(\theta, \tau) = \int s(t + \tau/2)s^*(t - \tau/2)e^{-j\theta t}dt. \tag{8}$$

Substituting (4) for $s(t)$ in this equation yields

$$\chi(\theta, \tau) = \int \int \int A(t + \tau/2, \omega_1)A^*(t - \tau/2, \omega_2)$$
$$\times e^{j(\omega_1(t+\tau/2)-\omega_2(t-\tau/2)-\theta t)}dZ(\omega_1)dZ^*(\omega_2)dt. \tag{9}$$

Following [13], $A(t, \omega)$ is assumed to be a slowly-varying envelope of a complex exponential, such that $A(t + \tau/2, \omega) \approx A(t, \omega)$. Taking the expectation of the ambiguity function and applying the orthogonality of $dZ(\omega)$ yields

$$E\left[\chi(\theta, \tau)\right] \approx \int \int |A(t, \omega)|^2 e^{j(\omega\tau - \theta t)} d\omega dt. \tag{10}$$

Since $|A(t, \omega)|^2 = P(t, \omega)$, this equation provides an approximate relationship between the expected value of the ambiguity function of the signal and the Fourier transform of the TFD. Since the statistics of the signal are presumed to be unknown, the ambiguity function itself is used in place of its expected value.

The task now is to find a positive distribution which satisfies the time and frequency marginals, and whose two-dimensional Fourier transform yields the ambiguity function of the signal. The immediately obvious problem is that no such distribution exists! Only the Wigner distribution, which generally not everywhere positive, satisfies this relationship. One way around this impass is to find the positive TFD whose Fourier transform equals the ambiguity function over a subset of the ambiguity plane. A second approach is to minimize the squared error between the transform of $P(t, \omega)$ and the ambiguity function subject to positivity and the marginals. Substituting $P(t, \omega)$ for $|A(t, \omega)|^2$ in (10) yields the following problem statement:

$$\min_{P} \int \int \left| \int \int P(t, \omega) e^{j(\omega\tau - \theta t)} dt d\omega - \chi(\theta, \tau) \right|^2 d\theta d\tau \tag{11}$$

$$\text{subject to} \quad P(t, \omega) \geq 0,$$

$$\int P(t, \omega) d\omega = |s(t)|^2,$$

$$\int P(t, \omega) dt = |S(\omega)|^2.$$

This is simply a quadratic programming problem with linear equality and inequality constraints.

The above problem formulation weights all values of the ambiguity function equally. However, it may be desirable to vary the weighting of the ambiguity function over the (θ, τ)-plane. Such a weighting is particularly useful in cases when prior knowledge of the signal characteristics is available, as in radar. The optimization problem is then transformed to a weighted least-squares problem by including a non-negative weighting function $w(\theta, \tau)$ in the minimization:

$$\min_{P} \int \int w(\theta, \tau) \left| \int \int P(t, \omega) e^{j(\omega\tau - \theta t)} dt d\omega - \chi(\theta, \tau) \right|^2 d\theta d\tau. \tag{12}$$

The constraints for this problem are unchanged. A subset of the ambiguity plane can be ignored, forming a "don't care" region, by setting $w(\theta, \tau)$ to zero. The weighting function can also be used as a lowpass filter, which allows the user to compute the TFD at a lower resolution in time and/or frequency.

B. Quadratic Programming

Quadratic programming (QP), and nonlinear programming in general, is a standard problem in optimization theory for which a large literature and numerous books exist (*e.g.* [1,9,10]). The usual form for a linearly-constrained QP problem is:

$$\text{Minimize} \quad \tfrac{1}{2}\mathbf{x}'\mathbf{H}\mathbf{x} - \mathbf{c}'\mathbf{x}$$
$$\text{subject to} \quad \mathbf{A}\mathbf{x} \le \mathbf{b}$$
$$\mathbf{x} \ge 0. \tag{13}$$

Equality constraints such as the marginals can be written as two inequality constraints: $\mathbf{A}\mathbf{x} \le \mathbf{b}$ and $\mathbf{A}\mathbf{x} \ge \mathbf{b}$. By discretizing the integrals and expanding the magnitude-squared term, (12) is easily written in the form of a QP problem:

$$\min_{P} \Bigg[\sum_{\theta,\tau} \sum_{t_1,\omega_1} \sum_{t_2,\omega_2} w(\theta,\tau) P(t_1,\omega_1) P(t_2,\omega_2) e^{j[(\omega_1-\omega_2)\tau - \theta(t_1-t_2)]}$$

$$- 2\Re \sum_{\theta,\tau} \sum_{t,\omega} w(\theta,\tau) P(t,\omega) \chi(\theta,\tau) e^{j(\omega\tau-\theta t)} + \sum_{\theta,\tau} |\chi(\theta,\tau)|^2 \Bigg] \tag{14}$$

where $\Re$ denotes the real part. Several simplifications can be made at this point. The last term is independent of $P(t,\omega)$ and can be dropped from the minimization. Secondly, the weighting function is equivalent to a kernel of a bilinear time-frequency distribution $C(t,\omega; \Phi_w)$ computed with the kernel $\Phi_w(t,\omega) = \mathcal{F}_{\theta,\tau}\{w(\theta,\tau)\}$, where $\mathcal{F}$ denotes a Fourier transform. The second term thus simplifies to:

$$\sum_{\theta,\tau} \sum_{t,\omega} w(\theta,\tau) P(t,\omega) \chi(\theta,\tau) e^{j(\omega\tau-\theta t)} = \sum_{t,\omega} P(t,\omega) C(t,\omega; \Phi_w). \tag{15}$$

The weighting function for the least-squares minimization must be real and positive; choosing a symmetric function will result in a real $C(t,\omega; \Phi_w)$. When $C(t,\omega; \Phi_w)$ is complex, only the real part will be preserved in (14), since $P(t,\omega)$ is real. Denoting the real part of $C(t,\omega; \Phi_w)$ by $C_r(t,\omega; \Phi_w)$ (noting that the two are equivalent for $w(\theta,\tau)$ symmetric), (14) reduces to

$$\min_{P} \Bigg[\sum_{t_1,\omega_1} \sum_{t_2,\omega_2} P(t_1,\omega_1) P(t_2,\omega_2) \Phi_w(t_1-t_2, \omega_1-\omega_2)$$

$$- 2 \sum_{t,\omega} P(t,\omega) C_r(t,\omega; \Phi_w) \Bigg]. \tag{16}$$

Dividing through by two yields a form equivalent to (13), with $\mathbf{x} \equiv P(t,\omega)$ (written out in vector form), $\mathbf{H} \equiv \Phi_w$ and $\mathbf{c} \equiv C_r(t,\omega; \Phi_w)$. $\mathbf{A}$ and $\mathbf{b}$ correspond to the marginal constraints on $P(t,\omega)$. This problem can then be solved with standard QP algorithms, such as that given in [9] (pp. 299–300).

IV. Discussion

As discussed before, least-squares computation of positive TFDs is not new. The general approach just described is distinct from these earlier methods in two ways. First, the derived algorithm is motivated by and based upon a statistical formulation of positive TFDs. In contrast, the earlier methods find a positive TFD which is "close" to a specified bilinear time-frequency representation. An interesting aspect of the QP method is how similar it is to the LS approaches, despite the difference in problem formulation. The only difference between the problem statements is the operator Φ_w in the quadratic term. Since both methods are completely defined by the kernel Φ_w, a comparison of the two methods for identical kernels should be interesting.

While TFD estimation is well-defined in a least-squares optimization framework, entropy methods are still preferred. For distribution estimation, including spectral densities, the appropriate objective function has been shown to be maximum entropy, or more generally, minimum cross-entropy [15,4]. Maximum entropy estimation provides the smoothest spectrum satisfying the constraints, and can be thought of as "making the fewest assumptions" about the data. For the constraints derived above, the optimization problem is written:

$$\min_{P} \int \int P(t, \omega) \log P(t, \omega) dt d\omega \tag{17}$$

$$\text{subject to} \quad \left| \int \int P(t, \omega) e^{j(\omega\tau - \theta t)} dt d\omega - \chi(\theta, \tau) \right| \leq \epsilon,$$

where ϵ is a small tolerance. Positivity and the marginals would also be included as constraints.

The advantage of using least-squares to compute TFDs instead of maximum entropy is primarily speed. Quadratic programming is an easier problem. The efficiency stems from the use of Parseval's relation in the derivation: the total squared error in the ambiguity domain is identical to that in the time-frequency plane. As a result, no Fourier transforms need to be computed during the iterative procedure. Such is not the case for entropy optimization; the constraints (in the ambiguity plane) need to be evaluated explicitly at each iteration. In addition, a linearly-constrained QP problem with N variables will converge in N iterations [1].

There are several refinements to the procedure that can be employed to improve the speed and quality of the result. The simplest is subsampling the TFD. Normally a TFD is computed at some lower time resolution, with frequency resolution determined by the extent of the kernel in τ. By limiting the weighting function to a finite extent in (θ, τ), $\Phi_w(t, \omega)$ and $C(t, \omega; \Phi_w)$ may be subsampled at the corresponding resolution; $w(\theta, \tau)$ is essentially acting as a decimating lowpass filter. $P(t, \omega)$ is then computed on this reduced domain. Since the problem is entirely defined in (t, ω), it should also be possible to design an efficient on-line algorithm.

Further improvements can be made by using an adaptive weighting function. For example, it may be advantageous to more heavily weight regions of the ambiguity plane where the ambiguity function of the signal is large. One way of doing this is to use the squared magnitude of the signal ambiguity function as the weighting function. Since the magnitude-square of an ambiguity function is its own two-dimensional Fourier transform, the corresponding kernel $\Phi_w(t, \omega)$ is just $|\chi(\omega, t)|^2$. A short-time ambiguity function can be used for an on-line adaptive algorithm.

References

1. M. Bazaraa, H. Sherali, and C. Shetty, *Nonlinear Programming: Theory and Algorithms*, New York, N. Y.: John Wiley and Sons, 2nd edition, 1993.

2. L. Cohen, *Time-Frequency Analysis*, Prentice-Hall, 1995.

3. L. Cohen, and T. Posch, "Positive Time-Frequency Distributions," *IEEE Transactions on Acoustics, Speech, and Signal Processing*, vol. 33, no. 1, 1985, pp. 31–38.

4. I. Csiszar, "Why Least-Squares and Maximum Entropy?—An Axiomatic Approach," *Ann. Stats.*, vol. 19, 1991, pp. 2033–2066.

5. J. Fonollosa, "Positive Time-Frequency Distributions Based on Joint Marginal Constraints," *IEEE Transactions on Signal Processing*, vol. 44, no. 8, 1996, pp. 2086–2091.

6. S. Kikkawa, and H. Tohyama, "A Least Squares Method for Generating Positive Time-Frequency Distributions," unpublished manuscript.

7. P. Loughlin, J. Pitton, and L. Atlas, "An Information-Theoretic Approach to Positive Time-Frequency Distributions," *ICASSP-92*, vol. V, 1992, pp. 125–128.

8. P. Loughlin, J. Pitton, and L. Atlas, "Construction of Positive Time Frequency Distributions," *IEEE Transactions on Signal Processing*, vol. 42, no. 10, 1994, pp. 2697–2705.

9. D. G. Luenberger, *Optimization by Vector Space Methods*, Wiley, 1969.

10. D. G. Luenberger, *Linear and Nonlinear Programming*, Addison-Wesley, 2nd ed., 1984.

11. J. Pitton, L. Atlas, and P. Loughlin, "Applications of Positive Time-Frequency Distributions to Speech Processing," *IEEE Trans. on Speech and Audio Proc.*, vol. 2, no. 4, 1994, pp. 554–566.

12. J. Pitton, P. Loughlin, and L. Atlas, "Positive Time-Frequency Distributions Via Maximum Entropy Deconvolution of the Evolutionary Spectrum," *ICASSP-93*, vol. IV, 1993, pp. 436–439.

13. M. Priestley, *Spectral Analysis of Time Series*, London: Academic Press, 1981.

14. T. Sang, W. Williams, and J. O'Neill, "An Algorithm for Positive Time-Frequency Distributions," *Proc. IEEE-SP Symp. Time-Freq./Time-Scale Analysis*, 1996, pp. 165–168.

15. J. Shore, and R. Johnson, "Axiomatic Derivation of the Principle of Maximum Entropy and the Principle of Minimum Cross-Entropy," *IEEE T-IT*, vol. 26, no. 1, 1980, pp. 26–37.

16. H. Tohyama, S. Kikkawa, and K. Ohara, "On an Optimum Positive Time-Frequency Distribution by the Least Squares Method (in Japanese)," *The Transactions of the Institute of Electronics, Information and Communication Engineers*, vol. J75-A1, no. 3, 1992, pp. 661–663.

Multidimensional Systems and Signal Processing, 9, 447–452 (1998)
© 1998 Kluwer Academic Publishers, Boston. Manufactured in The Netherlands.

On the Computation of Optical Flow using the 3-D Gabor Transform

TODD R. REED   trreed@ucdavis.edu
Department of Electrical and Computer Engineering, University of California, Davis, 3175 Engineering II

Received August 2, 1996; Accepted January 6, 1998

Abstract. The motion of brightness patterns in an image sequence (optical flow) is most intuitively considered as a spatiotemporal phenomenon. It has been shown, however, that motion has a characteristic signature in the spatiotemporal-frequency (Fourier) domain. This fact can be exploited for the computation of optical flow. However, for cases which involve a number of regions in a sequence with different motions, as in scenes with one or more objects moving against a stationary or moving background, the global nature of the Fourier transform makes it unsuitable for this task. The signatures of the different motions cannot be resolved in the Fourier domain, nor associated with their respective regions in the image sequence. Local frequency representations provide a means to address this problem. In this paper, we consider the application of a 3-D version of the widely used Gabor transform to the computation of optical flow.

Key Words: motion estimation, optical flow, Gabor transform, local frequency representations

1. Introduction

The majority of work in motion analysis focuses on spatiotemporal (st) methods, such as gradient-based techniques ([1] being a classical example) and the various block-matching methods popular in video compression applications. However, motion can also be characterized in the spatiotemporal-frequency (stf) domain. As shown in [2], an image undergoing uniform translational motion occupies a tilted plane in the stf domain. That is, an image sequence consisting of an image $f(x, y)$ moving with constant velocity $\mathbf{r} = [r_x, r_y]^T$,

$$f_\mathbf{r}(x, y, t) = f(x - r_x t, y - r_y t, t) \tag{1}$$

can be shown to have the Fourier transform

$$\mathcal{F}(f(x - r_x t, y - r_y t, t)) = F(u, v, w + r_x u + r_y v) \tag{2}$$

where $F(u, v)$ is the Fourier transform of $f(x, y)$, and u, v, and w are the spatial and temporal frequency variables. If the image was originally static, then

$$w = -(r_x u + r_y v). \tag{3}$$

Geometrically, the image motion changes the static image transform (which lies in the (u, v) plane) into a planar spectrum with slope $-r_x$ in the (u, w) plane and $-r_y$ in the (v, w) plane. For more complex motion, the energy in the stf domain is no longer planar, but still exhibits a well defined structure (as shown for translations with piecewise cubic trajectories in [3]).

If the sequence to be analyzed has only a single motion component, or if motion detection alone is required, the Fourier transform can be used as the basis for motion analysis, e.g.

115

[4]–[6]. However, due to the global nature of the Fourier transform, it cannot be used to determine the location of the object in motion. It is also poorly suited for cases in which multiple motions exist (i.e., when the scene of interest consists of more than one object moving independently), since the signatures of the different motions are difficult to separate in the Fourier domain. To identify the locations and motions of objects, frequency analysis localized to the neighborhoods of the objects is required. Windowed Fourier analysis has been proposed for such cases [7]. However, the accuracy of a motion analysis method of this type is highly dependent on the resolution of the underlying transform, in both the st and stf domains. It is known that the windowed Fourier transform is not the best choice in this regard.

Alternative local frequency decompositions that might be considered include the Wigner distribution and the Gabor transform. The use of the Wigner distribution for this task was examined by Jacobson and Wechsler in [8]. In this paper, we consider the use of the Gabor transform.

2. The Gabor Representation

The Gabor representation was first introduced for time-frequency analysis by D. Gabor in 1946 [9]. The functions upon which the representation is based are each formed by the product of a Gaussian window (centered at a particular point in time) and a complex exponential (which establishes the center of the function in the frequency domain). It has been shown that under certain conditions a complete basis can be formed using these functions, resulting in an invertible transform. The 1-D and 2-D Gabor transforms have proven very popular in a variety of signal and image processing applications.

If we consider an image sequence to be a 3-dimensional (spatiotemporal) volume of data, the sequence can be represented via the 3-D Gabor transform as the weighted sum of 3-D Gabor functions of the form:

$$g(x, y, t) = \hat{g}(x, y, t)e^{j(u_0(x-x_0)+v_0(y-y_0)+w_0(t-t_0))}, \tag{4}$$

where

$$\hat{g}(x, y, t) = \frac{1}{(2\pi)^{3/2}\sigma_x\sigma_y\sigma_t} \, e^{-[((x-x_0)/\sigma_x)^2+((y-y_0)/\sigma_y)^2+((t-t_0)/\sigma t)^2]/2}$$

is a 3-D Gaussian function, σ_x, σ_y and σ_t determine the scale of the Gaussian along the respective axes, (x_0, y_0, t_0) is the center of the function in the st domain, and (u_0, v_0, w_0) is the center of support in the stf domain.

Because the Gabor transform is not orthogonal, it cannot be calculated by computing the inner products of the basis functions and the signal of interest (or, equivalently, by convolving with the basis functions and subsampling). Approaches to computing this transform have received a great deal of attention recently, with the result that a variety of methods are available. These methods vary in the restrictions placed on the basis set of the transform, and in computational efficiency. The most efficient (based on the Zak transform) is of the same computational complexity as the FFT. The method used in the experiment that follows is an extension of the algorithm first reported in [10].

116

Figure 1. The first and fifth frames from the test sequence.

3. The Computation of Optical Flow

Given the 3-D Gabor transform of an image sequence, the optical flow vector **r** can be estimated at each spatiotemporal location by finding the slope of the spectrum at that location. A simple approach to this problem is to form vectors of the u, v, and w coordinates of the basis functions that have significant energy for each point in the sequence at which basis functions are centered. Then from equation 3,

$$\mathbf{w} = -(r_x \mathbf{u} + r_y \mathbf{v}) = -\mathbf{A}\mathbf{r} \tag{5}$$

where $\mathbf{A} = (\mathbf{u}|\mathbf{v})$. An LMS estimate of the optical flow vector at a given point can then be found using the pseudoinverse of $\mathbf{A}$:

$$\mathbf{r} = -(\mathbf{A}^T \mathbf{A})^{-1} \mathbf{A}^T \mathbf{w} \tag{6}$$

4. Results

To demonstrate the method outlined above, we will use the simple test sequence shown in Figure 1. The sequence is 24 frames in length, with 256-by-256 pixels per frame. Each frame consists of two fields, both consisting of horizontal and vertical sinusoids with frequencies of $\pi/4$, one field above the other. The sinusoids are scaled and offset so that all pixel intensities fall in the range 0 to 255. As the sequence progresses, the top field moves to the left at 1 pixel/frame, while the bottom field moves to the right at the same rate. Fourier analysis of the sequence would yield (from equation 3) nonzero coefficients at $(u, v, w) = (0, 0, 0)$, $(0, \pi/4, 0)$, $(0, 7\pi/4, 0)$, $(\pi/4, 0, \pi/4)$, $(7\pi/4, 0, 7\pi/4)$, $(\pi/4, 0, 7\pi/4)$, and $(7\pi/4, 0, \pi/4)$, where the terms at $7\pi/4$ are due to the conjugate symmetry of the Fourier transform for real signals. Due to the global nature of the Fourier transform, the correspondence between these coefficients and their spatiotemporal locations is indeterminate.

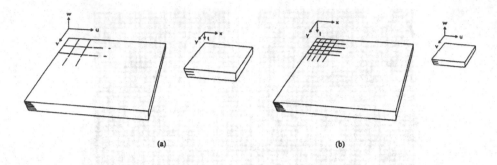

Figure 2. Two projections of the transform coefficients: (a) grouped by frequency; (b) grouped by location.

We next compute the Gabor transform of the two sequences, using a complete basis on 8 pixel centers in space-time, spaced $\pi/4$ apart in spatiotemporal-frequency, and with $\sigma_x = \sigma_y = \sigma_t = 1/8$. Because the resulting transform is 6-dimensional, it is necessary to project the coefficients onto 3 dimensions for display. Two useful projections are shown in Figure 2. Each subblock shown to the left in Figure 2a corresponds to a specific spatiotemporal frequency. Within each subblock, the coefficients are arranged in spatiotemporal registration with the input sequence, as shown in the subblock to the right in Figure 2a. In Figure 2b, each subblock on the left corresponds to a specific spatiotemporal location. Within each subblock, as shown to the right, coefficients are arranged according to their associated frequencies. The first projection is particularly useful for observing the degree to which a particular frequency is represented at specific points in the sequence. The second, which presents the data in the form of local spectra, is the one we will use in the computation of optical flow.

The magnitudes of the Gabor transform coefficients for the test sequence are shown in Figure 3. Coefficients with low magnitudes are rendered as translucent. From the first projection (Figure 3, left), it can be seen that the sequence exhibits energy at $(u, v, w) = (0, 0, 0)$, $(0, \pi/4, 0)$ and $(0, 7\pi/4, 0)$ at all spatial locations, at $(\pi/4, 0, \pi/4)$ and $(7\pi/4, 0, 7\pi/4)$ for the "top" of the sequence (which moves to the left) and at $(\pi/4, 0, 7\pi/4)$ and $(7\pi/4, 0, \pi/4)$ for the "bottom" (which moves to the right). The terms at $7\pi/4$ are due to the conjugate symmetry of the Gabor transform for real signals.

Computing the slope of the plane which best fits each local spectrum (shown in Figure 3, right), the optical flow map shown in Figure 4 results for the point in time between frames 12 and 13 of the sequence. Note that the optical flow estimates are located between frames because the basis functions used in the transform are centered between frames. The velocity estimates shown in Figure 4 are correct in both direction and magnitude (1 pixel/frame to the left and right, respectively). Similar results are obtained between frames 4 and 5, and frames 20 and 21.

Figure 3. The magnitudes of the coefficients for the test sequence grouped by frequency (left) and location (right).

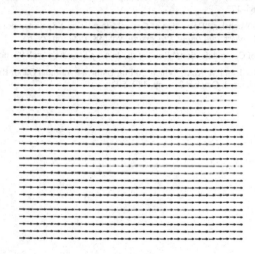

Figure 4. The optical flow computed between frames 12 and 13 of the sequence.

5. Conclusions and Future Work

In this paper, we have shown that optical flow can be computed via the Gabor transform. Unlike Fourier-based methods, this approach allows sequences with multiple motion components to be successfully analyzed, with the spatiotemporal location of each component identified. Future work includes the evaluation of this method on sequences of natural scenes, and the extension of this approach to produce dense (pixel and subpixel resolution) motion estimates.

References

1. B. K. P. Horn, and B. G. Schunk, "Determining Optical Flow," *Artificial Intelligence*, vol. 17, no. 1–3, 1981, pp. 185–203.

2. A. B. Watson, and A. J. Ahumada, "A Look at Motion in the Frequency Domain," *SIGGRAPH/SIGART Interdisciplinary Workshop MOTION: Representation and Perception*, Toronto, Canada, April 4–6 1983, pp. 1–10.

3. G. Cortelazzo, and M. Balanza, "Frequency Domain Analysis of Translations with Piecewise Cubic Trajectories," *IEEE Trans. PAMI*, vol. 15, no. 4, 1993, pp. 411–416.

4. H. Gafni, and Y. Y. Zeevi, "A Model for Separation of Spatial and Temporal Information in the Visual System," *Biol. Cybern.*, vol. 28, 1977, pp. 73–82.

5. B. Porat, and B. Friedlander, "A Frequency Domain Algorithm for Multiframe Detection and Estimation of Dim Targets," *IEEE Trans. PAMI*, vol. 12, no. 4, 1990, pp. 398–401.

6. A. Kojima, N. Sakurai, and J. Kishigami, "Motion Detection Using 3D-FFT Spectrum," *Proc. of IEEE ICASSP '93*, Minneapolis, Minnesota, April 27–30, pp. 213–216.

7. H. Gafni, and Y. Y. Zeevi, "A Model for Processing of Movement in the Visual System," *Biol. Cybern.*, vol. 32, 1979, pp. 165–173.

8. H. Jacobson, and H. Wechsler, "Derivation of Optical Flow Using a Spatiotemporal-Frequency Approach," *Computer Vision, Graphics, and Image Processing*, vol. 38, 1987, pp. 29–65.

9. D. Gabor, "Theory of Communication," *Proc. of the Inst. of Elec. Eng.*, vol. 93, no. 26, 1946, pp. 429–457.

10. T. Ebrahimi, T. R. Reed, and M. Kunt, "Video Coding Using a Pyramidal Gabor Expansion," *Proc. of VCIP '90*, Lausanne, Switzerland, October 2–4, pp. 489–502.

Multidimensional Systems and Signal Processing, 9, 453–458 (1998)
© 1998 Kluwer Academic Publishers, Boston. Manufactured in The Netherlands.

Evolutionary Maximum Entropy Spectral Estimation and Heart Rate Variability Analysis

S. I. SHAH
Department of Electrical Engineering, 348 Benedum Hall, University of Pittsburgh, Pittsburgh, PA 15261

L. F. CHAPARRO
Department of Electrical Engineering, 348 Benedum Hall, University of Pittsburgh, Pittsburgh, PA 15261

A. EL-JAROUDI
Department of Electrical Engineering, 348 Benedum Hall, University of Pittsburgh, Pittsburgh, PA 15261

J. M. FURMAN
Department of Otolaryngology, 200 Lothrop Hall, University of Pittsburgh, Pittsburgh, PA 15261

Received August 9, 1996; Revised February 12, 1997

Abstract. Spectral analysis has been used extensively in heart rate variability (HRV) studies. The spectral content of HRV signals is useful in assessing the status of the autonomic nervous system. Although most of the HRV studies assume stationarity, the statistics of HRV signals change with time due to transients caused by physiological phenomena. Therefore, the use of time-frequency analysis to estimate the time-dependent spectrum of these non-stationary signals is of great importance. Recently, the spectrogram, the Wigner distribution, and the evolutionary periodogram have been used to analyze HRV signals. In this paper, we propose the application of the evolutionary maximum entropy (EME) spectral analysis to HRV signals. The EME spectral analysis is based on the maximum entropy method for stationary processes and the evolutionary spectral theory. It consists of finding an EME spectrum that matches the Fourier coefficients of the evolutionary spectrum. The spectral parameters are efficiently calculated by means of the Levinson algorithm. The EME spectral estimator provides very good time-frequency resolution, sidelobe reduction and parametric modeling of the evolutionary spectrum. With the help of real HRV signals we show the superior performance of the EME over the earlier methods.

Key Words: non-stationary time series, evolutionary spectral theory, spectral analysis of heart rate variability, cardiovascular time series

I. Introduction

The spectral analysis of heart rate variability (HRV) signals permits the study of autonomic influences on the heart rate. In performing this analysis, the HRV signal is extracted from the electrocardiogram (ECG) signal by means of one of several methods [1]. Under the assumption of stationarity, power spectral analysis is performed on the HRV signal. However, this signal is, in general, non-stationary due to physiological phenomena, such as physical and emotional stresses, or to postural changes causing rapid transients [2] in the signal. This non-stationarity in the signal is characterized by a time-dependent spectrum and, as such, time-frequency analysis [3] may be used to estimate it.

Time-frequency analysis of HRV signals has been performed with methods such as the spectrogram [2], the Wigner distribution [4] and the evolutionary periodogram [5]. While the time-frequency resolution of the spectrogram is badly affected by the windowing, the Wigner distribution cannot guarantee positivity of the estimates and displays cross terms.

121

The evolutionary periodogram, on the other hand, guarantees positivity and displays no cross terms, and provides some improvement on the time-frequency resolution over the spectrogram. In this paper, we propose the analysis of HRV signals using evolutionary maximum entropy (EME) [6]. EME spectral analysis is based on the maximum entropy method for stationary processes and evolutionary spectral theory. It provides good time-frequency resolution, sidelobe reduction and parametric modeling of the evolutionary spectrum.

II. HRV Signal Extraction

In a typical electrocardiogram (ECG) signal, the R-wave (peak of the ECG) is generally considered the component of interest in heart rate variability studies. The intervals between R-waves are not regular due to different autonomic influences, and as such these values constitute a point process. The IHR signal is a step-wise continuous signal obtained from the point process in the following way. For each interval, the heart rate is computed and taken as the amplitude of the instantaneous heart rate (IHR) signal for the duration of that interval. The spectrum of the IHR signal is calculated to obtain a spectral description of the HRV. Unfortunately, the discontinuities in the IHR signal at the interval boundaries result in high frequency components in the spectrum that cannot be justified physiologically. To avoid this problem, we interpolate the intermediate values in the IHR using a cubic spline [7], instead of holding the current heart rate constant until the next beat. This results in a continuous and smooth signal whose time-dependent spectrum we then calculate using EME spectral analysis.

III. Evolutionary Maximum Entropy Spectral Analysis

A non-stationary signal $x(n)$ can be represented as the output of a linear time-varying (LTV) system with white noise as input, yielding

$$x(n) = \int_{-\pi}^{\pi} H(n, \omega) e^{j\omega n} dZ(\omega) \tag{1}$$

where $Z(\omega)$ is a process with orthogonal increments, and $H(n, \omega)$ is the generalized transfer function of the LTV system evaluated on the unit circle. Equation (1) is the Wold-Cramer representation [9] of $x(n)$, according to which $x(n)$ is viewed as a sum of sinusoids with random and time-varying amplitudes and phases. The Wold-Cramer evolutionary spectrum [9,10] of $x(n)$ is accordingly defined as

$$S(n, \omega) = |H(n, \omega)|^2. \tag{2}$$

The underlying principle for EME is to calculate, at each time n, a maximum entropy spectrum which is constrained to match time-dependent statistics calculated from the data. In other words, at each n, we maximize

$$\int_{-\pi}^{\pi} \ln S(n, \omega) d\omega \tag{3}$$

under the constraints that, for some value P_n,

$$\frac{1}{2\pi} \int_{-\pi}^{\pi} S(n, \omega) e^{j\omega\mu} d\omega = f(n, \mu) \qquad -P_n \leq \mu \leq P_n \tag{4}$$

and that $S(n, \omega) > 0$, where $f(n, \mu)$ are the Fourier coefficients of the evolutionary spectrum. It is noteworthy that the constraints in (4) cannot be written with respect to the time-dependent autocorrelations of the signal since the evolutionary spectrum and the time-dependent autocorrelations do not form a Fourier transform pair [6].

At a given time, maximizing (3) coincides with the method of maximum entropy estimation for the stationary case [11]. Therefore, at each n, the spectral solution is that of an autoregressive model of order P_n,

$$S_{EME}(n, \omega) = \frac{\varepsilon(n)}{A(n, \omega) A^*(n, \omega)} \tag{5}$$

where $A(n, \omega) = 1 + \sum_{k=1}^{P_n} a(n, k) e^{-j\omega k}$, and $\varepsilon(n)$ is a spectral normalization factor. The $\{a(n, k)\}$ coefficients are obtained by solving

$$\sum_{k=1}^{P_n} a(n, k) f(n, k - \mu) = -f(n, \mu) \qquad 0 \leq \mu \leq P_n \tag{6}$$

which is efficiently performed by means of the Levinson algorithm. Notice that the above is not equivalent to the "frozen-time" spectral estimation suggested by some authors and which, as clarified by Melard [10], does not relate to the evolutionary spectrum.

The solution of the above normal equations requires that we have estimates for the $\{f(., .)\}$ function. A family of estimators for $f(., .)$ can be obtained using various evolutionary spectral estimators. We select an estimator $\hat{f}(., .)$ that is obtained from the evolutionary periodogram [8], and can be written as

$$\hat{f}(n, \mu) = \frac{N}{M} \sum_{m=K_1}^{K_2} w_n(m + \mu) x(m + \mu) w_n^*(m) x^*(m)$$
$$K_1 = 0, K_2 = N - 1 - \mu, \text{ if } \mu \geq 0$$
$$K_1 = -\mu, K_2 = N - 1, \text{ if } \mu < 0 \tag{7}$$

where $w_n(m)$ is a time-varying window given by

$$w_n(m) = \sum_{i=0}^{M-1} \beta_i^*(n) \beta_i(m) \tag{8}$$

for a set of orthonormal functions $\{\beta_i(.)\}$. It is important to note that by setting the expansion order $M = 1$, and $\beta_0(n) = \frac{1}{\sqrt{N}}$, the EME reduces to the classical maximum entropy method.

In practice, the EME analysis depends on the expansion order M, and on the constraints order P_n. The parameter $M \geq 1$ indicates the degree of non-stationarity of the signal. The order P_n can be chosen just as in the stationary case [11], given that at each time the method is similar to classical maximum entropy. The time-frequency resolution of the EME estimates depends on the choices of M and P_n. Also, the choice of the orthonormal functions $\{\beta_i\}$ in (8) depends on the application. We have found the Fourier and the Legendre discrete polynomials very useful in our analysis (See reference [8] for more details).

IV. Example

As an example to illustrate the performance of EME in HRV analysis, we use a 180-second ECG signal from a cat, sampled at 1000 Hz. The data are collected such that the head of the animal is kept level for 60 seconds, then its head is pitched nose-up by 50 degrees for 60 seconds and finally the head is kept flat for 60 seconds. The resulting HRV signal is obtained using a cubic spline interpolation. To remove small drift in the mean HRV signal we subtract from it an estimate of its time-varying mean. The resulting HRV signal between 47.25 and 79.25 seconds is shown in Fig. 1(a). Fig. 1(b) displays the spectrogram of the HRV signal, computed by using an 8-second Hamming window. While the spectrogram succeeds at demonstrating the HRV signal is indeed composed of narrowband components (the dark areas in the figure), it also displays broad smearing in time and frequency. Fig. 1(c) shows the Wigner distribution of the HRV signal. The distribution displays excellent localization of the energy in time and frequency, however it also shows too many spurious components due to the effects of cross terms. The evolutionary periodogram estimate with 9 Fourier functions is shown in Fig. 1(d). While a little sharper than the spectrogram, the evolutionary periodogram estimate still suffers from smearing. Fig. 1(e) shows the EME estimate with a model order P_n of 60. The result clearly shows the narrowband components (the darkest areas in the figure) of the HRV signal and their trajectories in time-frequency. It also clearly displays the discontinuity at the 60th second due to the change in the head position of the animal. The EME estimate has better frequency resolution than the spectrogram and the evolutionary periodogram and is free of the cross terms found in the Wigner distribution.

V. Conclusions

In this paper we have presented a new method to estimate the time-dependent spectrum of non-stationary signals. The EME has the advantage of being based on well-known stationary spectral methods, and improves on previous evolutionary spectral estimators. We have also explored the application of the EME spectral estimator to the HRV signal. The EME provides a better estimate of the changes in the HRV signal than short-time spectral analysis and other currently available time-frequency methods. It also provides a reduction in the side-lobe behavior, higher time-frequency resolution, and can be used in the parametric modeling of the HRV signal.

Acknowledgements

The authors thank Dr. Bill Yates of the University of Pittsburgh Medical Center, for providing the ECG data used in the analysis.

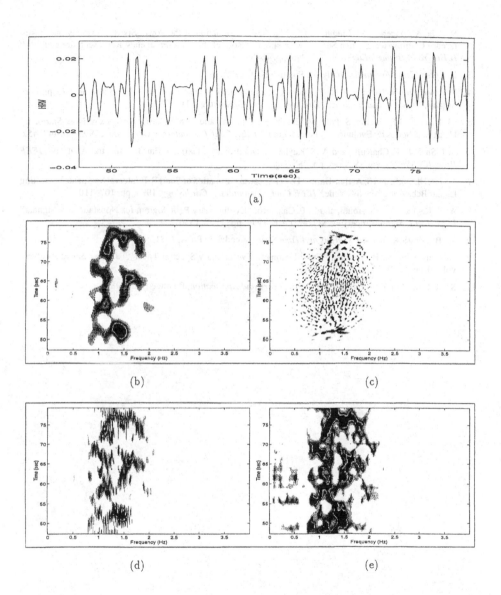

Figure 1. (a) HRV signal,(b) Spectrogram, (c) Wigner distribution, (d) Evolutionary Periodogram and (e) Evolutionary Maximum Entropy.

References

1. R. D. Berger, S. Akserlord, D. Gordon, and R. J. Cohen, "An Efficient Algorithm for Spectral Analysis of Heart Rate Variability," *IEEE Trans. Biomed. Eng.*, vol. 33, 1986, pp. 900–904.

2. S. Pola, A. Macerata, M. Emdin, and C. Marchesi, "Estimation of the Power Spectral Density in Nonstationary Cardiovascular Time Series: Assessing the Role of the Time-Frequency Representations (TFR)," *IEEE Trans. Biomed. Eng.*, vol. 43, 1996, pp. 46–59.

3. L. Cohen, *Time-Frequency Analysis*, Prentice Hall, 1995.

4. P. Novak and V. Novak, "Time-Frequency Mapping of the Heart Rate, Blood Pressure and Respiratory Signals," *Medical and Biological Engineering, and Computing*, Mar. 1993.

5. M. Niccolai, M. Varanin, S. Pola, M. Emdin, M. Cipriani, and C. Marchesi, "Analysis of Non–Stationary Heart Rate Series by Evolutionary Periodogram," *IEEE Conf. Computers in Cardiology*, 1995, pp. 449–452.

6. S. I. Shah, L. F. Chaparro, and A. S. Kayhan, "Evolutionary Maximum Entropy Spectral Analysis," *IEEE Proc. ICASSP'94*, vol. IV, 1994, pp. 285–288.

7. T. Harel, I. Gath, and S. A. Ben-Haim, "Power Spectral Estimation of Short R-R Interval Segments Presenting Linear Behavior of the SA Node," *IEEE Conf. Computers in Cardiology*, 1995, pp. 107–110.

8. A. S. Kayhan, A. El-Jaroudi, and L. F. Chaparro, "Evolutionary Periodogram for Nonstationary Signals," *IEEE Trans. Signal Processing*, vol. 42, no. 6, 1994, pp. 1527–1536.

9. M. B. Priestley, *Spectral Analysis and Time Series*, Academic Press, 1981.

10. G. Melard and A. H. Schutter, "Contributions to Evolutionary Spectral Theory," *J. Time Series Analysis*, vol. 10, no. 1, 1989, pp. 41–63.

11. S. M. Kay, *Modern Spectral Estimation: Theory and Application*, Prentice Hall, 1988.

Multidimensional Systems and Signal Processing, 9, 459–463 (1998)
© 1998 Kluwer Academic Publishers, Boston. Manufactured in The Netherlands.

Optimal Reconstruction of Natural Images by Small Sets of Gabor Filters

J. H. VAN DEEMTER   deemter@rc.service.rug.nl
Center for High Performance Computing, University of Groningen, The Netherlands

G. CRISTÓBAL   gabriel@optica.csic.es
Instituto de Optica (CSIC), Serrano 121, 28006 Madrid, Spain

Received September 25, 1996; Accepted March 15, 1997

Abstract. Images can be reconstructed after being filtered by a Gaussian and a few Gabor filters. Several search methods for the filter parameters for a (near) optimal reconstruction are examined. At first, the search is performed on a 1-D signal which satisfies the radial spectrum of the average of natural images. Later, these results are extended and applied to a 2-D example. It is shown that the resulting parameters are better than those of some existing models.

Key Words: image reconstruction, aliasing, Gabor filters, image quality

1. Introduction

Images can be sampled by sets of different filters to obtain non-orthogonal, redundant and therefore more robust image representations. Such filter sets can enhance several features of an image, e.g. edges [6], multiscale aspects [4,2], or the orientation distribution [8]. These features can be used to improve the image quality, its understanding, e.g., by allowing for stability measures of edges in scale space [2], or the suppression of noise [8,4]. Such strategies not only proved to be beneficial in technical applications, but are also found in biological visual systems. In the retinas of mammals, ganglion cells have been found with a large variety of receptive field sizes (often modeled by the Laplacian of Gaussians, or "Mexican hat" functions [6]) and in the primary visual cortex simple cells of many different orientations, wavelengths and bandwidths abound (often modeled by 2-D Gabor functions [3,5]). It must be emphasized that the filter set should be as complete as possible, since it might be desirable to reconstruct the image from the filter responses [7]. Such a complete set can generally be obtained by using huge numbers of filters which can be described by an almost continuous range of parameters. Although this approach could mimic the visual systems of mammals, it may cause practical problems in artificial systems. Here, it would be preferable to obtain a limited set of filters from which the image can be reconstructed optimally. In this paper we discuss several aspects of finding an near optimal set of such filters. We will restrict ourselves to the use of one lowpass filter (a Gaussian) and a few bandpass filters (Gabors).

127

2.   Filter Set Selection

To reconstruct a natural image, it is first convolved with a Gaussian and a few Gabor filters. These filtered outputs are then linearly combined to obtain reconstruction. Both the Gaussian and Gabor filters are implemented in the frequency domain as Gaussians, where the lowpass Gaussian filter is centred around its central frequency $f_0 = 0$ and the bandpass Gaussians are centred around $\pm f_0 \neq 0$.

For the reconstruction we first compute the FFT of the image and then multiply the resulting spectrum with each of the filters of the filterset. Then, the results of these multiplications are summed for each f and the final outcome is transformed back to the space domain. With a small filter set consisting of a few lowpass and bandpass Gaussians it is not possible to construct a filter set that has value 1 for all the frequencies, i.e. does not change the image, but it is possible to approximate 1 for the most (essential) frequencies. In this case, the optimal filter parameters depend on the *amplitude* of the spectrum of the image. This means that each image requires another filter set which is specific for its spectrum. As a compromise we will construct our filter set for the average spectrum of a large set of natural images: $A(f) = cf^{-\alpha}$ where c and α are constants and f is the frequency [10]. In this study we will concentrate on an average image with $\alpha = 1$ and with no orientational preference. We will search for a small filter set that reconstructs such an image optimally. The assumed rotational invariance of the spectrum lets us reduce the amount of computations by only considering the 1-dimensional case. At the end we will generalize the results for 2-dimensional reconstructions.

2.1.   Boundary Conditions and Reconstruction Quality

The extent of the Gaussians in the frequency domain is infinite and the number of samples is limited. Therefore, they must be cut off and this introduces some errors at both the low end ($f = 0$) and the high end ($f = N/2$, aliasing) of the frequency domain. To reduce these errors, we will introduce the boundary condition that the fraction of the surface under a bandpass Gaussian lying outside the borders of the frequency domain may not exceed value p (e.g. 0.01).

Before starting a search for optimal reconstruction parameters the reconstruction quality should be defined. An appropriate definition is the Signal to Noise Ratio (SNR), which can be defined as the energy of the signal divided by the summed squared errors and can be used both in the space and the frequency domain

$$SNR = 10 \log_{10} \frac{\sum_{i=1}^{N} f_i^2}{\sum_{i=1}^{N} (f_i - f_i')^2} = 10 \log_{10} \frac{\sum_{i=1}^{N} F_i^2}{\sum_{i=1}^{N} (F_i - F_i')^2} \tag{1}$$

where f_i and f_i' depict the i-th samples of the original and reconstructed signal and F_i and F_i' are the i-th samples of the original and reconstructed spectrum.

3. Several Search Methods

The width of the lowpass Gaussian is set to a fixed value and therefore only the parameters of the bandpass Gaussians remain to be determined. In principle the magnitude, the central frequency f_0, and the width σ suffice for the specification of one Gaussian. In order to limit the degrees of freedom we only used one global scaling factor for the filters, which generally can be estimated reasonably well from the global statistics of the original and the reconstructed signal before scaling correction. For details see [9]. Hence, for *each* bandpass filter two parameters f_0 and σ remain. There are several methods for maximizing a function in a multidimensional space, but they strongly depend on its shape. We will discuss a few of these methods.

One way is first scanning the space with a coarse grid and then *progressively decreasing* the gridsize in the higher SNR regions. The problem with this method is that the filter parameters are strongly interdependent and that consequently the errors due to sampling next to the real maximum will grow progressively.

The well-known *gradient descent/ascent* method must also be rejected, since for our parameter space with boundary conditions too many local maxima exist in which the method gets stuck.

A solution that remains is the *brute search method.* Here the complete parameter space is scanned at an as small as possible grid size. This method does not guarantee to find the exact maximum, but can approach it. A disadvantage is the large computation costs which grow linearly with the number of grids of each extra parameter. We used this method for up to 2 Gabor filters. For a reasonable gridsize of 1 pixel (for both the f_0's and the σ's) this costs approximately $1.6 \; 10^3$ and $1.4 \; 10^6$ reconstructions for 1 and 2 Gabor filters, respectively. The reconstructions for the 2-Gabor case took about 30 minutes CPU time on a PC with a Pentium 90 processor after optimizing the program. A reconstruction with 3 Gabors would cost about $5.7 \; 10^8$ reconstructions which would be about 200 hours on the Pentium, and hence is a typical job for a supercomputer. To reduce the number of free parameters we decided to test 2 reconstruction models following two different band arrangement criteria(A and B) for the 3-Gabor case. For model A all the Gabors have the same width σ (equally linear bands) whereas in model B, the first Gabor has width σ, the second width 2σ and the third width 4σ (equal octave bands). In both cases σ and the 3 central frequencies are the 4 free parameters.

4. Experiments and Results

We used a 1-D signal of $N = 256$ samples with an average spectrum. With a small filter set we tried to find a configuration which results in a reconstruction with the largest SNR. For all the reconstructions we have set the lowpass residual to a Gauss with a small bandwidth ($\sigma = 2.5$), which allows for subsampling to $N = 16$ with little aliasing. Although a Gaussian with a larger extent may still satisfy the $p = 0.01$ condition, it covers too large a part of the signal spectrum eliminating the advantages of bandpass filtering of signals. In Table 1 the results of the brute search method are shown. Apart from the case $p = 0.01$ we

Table 1. Best reconstructions for 1 lowpass ($\sigma = 2.5$) and 1, 2 or 3 bandpass Gaussians. All the reconstructions are accompanied by the parameters $f_0 \pm \sigma$. For details see text.

	$p = 0.001$	$p = 0.01$	$p = 0.1$
1 Gabor	19.9 dB	23.2 dB	24.5 dB
	10 ± 2	10 ± 4	16 ± 11
2 Gabors	24.9 dB	27.8 dB	27.8 dB
	$7 \pm 2, 20 \pm 6$	$8 \pm 3, 31 \pm 13$	$9 \pm 5, 37 \pm 25$
3 Gabors, A		25.6 dB	
		$8 \pm 2.8, 16 \pm 2.8, 25 \pm 2.8$	
3 Gabors, B		27.5 dB	
		$6 \pm 1.8, 14 \pm 3.6, 30 \pm 7.2$	

also included the results for $p = 0.001$ and $p = 0.1$. For reference one should know that reconstruction with the lowpass Gaussian alone (with $\sigma = 2.5$) yields a SNR of 18.2 dB.

For the case of 3 Gabor filters we compared the two reconstruction models for $p = 0.01$. During the search, the σ's and f_0's for both models were varied by stepsize 0.2 and 1, respectively. For model A the $\sigma's$ ran from 0.2 to 24. In model B, the smallest ran from 0.2 to 6 and the larger σ's were set at 2σ and 4σ. Table 1 shows that the reconstructions obtained by the brute search method for the 2-Gabor reconstructions with 2 free f_0's and 2 free σ's are (slightly) better than each of the 3-Gabor reconstructions with 3 free f_0's but interdependent σ's. To see whether this result also holds for images with the spectrum differing from the average, we generalized the 1-D results to 2-D results by generating separable bandpass Gaussians. These can be written as a product of one Gaussian in the radial spatial frequency direction and one perpendicular to it. For the radial Gaussians the parameters from Table 1 are used, whereas for the perpendicular Gaussians a σ is taken of about 45^0, so that 4 channels per spatial frequency can roughly cover all orientations [1]. For the boundary condition $p = 0.01$ we compared the reconstruction quality of the *Lena* image for the 2-Gabor case and the 2 models A and B for the 3-Gabor case. The SNR's are 21.1, 18.8 and 19.8 dB, respectively, which shows that also in this case the 2-Gabor result is better than the two 3-Gabor models.

5.   Conclusions and Discussion

For weaker boundary conditions (i.e., for larger p-values) the SNR's are much larger. Therefore, if a reconstruction process is described with bandpass filters like Gabors, it is of crucial importance to mention the boundary conditions. Furthermore, in the 2-Gabor case, the bandpass Gaussian with the smallest f_0 has a smaller σ and f_0 than the bandpass Gaussian in the 1-Gabor case. This follows from the fact that this latter filter needs to find a compromise between covering the whole frequency domain and being positioned near $f = 0$ where the spectrum is larger.

The significance of an optimum found in a multidimensional parameter space is hard to determine. First, the optimum can only be found approximately. Second, the brute force method may completely miss the global optimum and find a local optimum instead. Fortunately, some trends towards an optimum could be seen when plotting the SNR's as a function of one of the search parameters and hence we don't expect the parameters for the optima to deviate too much from the parameters of the 'real' optima.

From the relatively good reconstruction results from the brute search method it may be concluded that the search for optimal reconstruction filter parameters can be useful and should be extended to larger numbers of filters in future, not restricted to Gabors only. The influence of the the size of the lowpass filter on the reconstruction results should also be incorporated in these investigations.

Acknowledgements

This research has been supported by the following grants: *PARAMONOP* CE-HCM Grant No. CHRX-CT94-0636; Grant No. 612.21.101 of the Dutch Science Foundation (NWO, SION); Spanish Commission for Science and Technology Grant No. TIC94-0849 and NATO Collaborative Grants Programme. We thank Dr. M. H. F. Wilkinson for his help in compressing the manuscript to within acceptable limits.

References

1. J. Bigun and J. M. H. du Buf, "N-Folded Symmetries by Complex Moments in Gabor Space and Their Application to Unsupervised Texture Segmentation," *IEEE Trans. Pattern Analysis Mach. Intell.*, vol. 16, no.1, 1994, pp. 80–87.

2. W. F. Bischof and T. Caelli, "Parsing Scale-Space and Spatial Stability Analysis," *Comput. Vision, Graphics and Image Processing*, vol. 42, 1988, pp. 192–205.

3. J. Daugman, "Two-Dimensional Spectral Analysis of Cortical Receptive Field Profiles," *Vision Res.*, vol. 20, 1980, pp. 847–856.

4. B. Liu, *Wavelet Phase Filter for Denoising in Tomographic Image Reconstruction*, PhD thesis, Illinois Inst. Tech., 1994.

5. S. Marcelja, "Mathematical Description of the Responses of Simple Cortical Cells," *J. Opt. Soc. Am. A*, vol. 70, no. 11, 1980, pp. 1297–1300.

6. D. Marr and E. Hildreth, "Theory of Edge Detection," *Proc. R. Soc. London B*, vol. 207, 1980, pp. 187–217.

7. R. Navarro and A. Tabernero, "Gaussian Wavelet Transform: Two Alternative Fast Implementations for Images," *Multidimen. Syst. Signal Process.*, vol. 2, 1991, pp. 421–436.

8. E. Peli, "Adaptive Enhancement Based on a Visual Model," *Opt. Eng.*, vol. 26, no.7, 1987, pp. 655–660.

9. J. H. Van Deemter and G. Cristóbal, "Signal Reconstruction with a Small Set of Gabor Filters," *SPIE Proceedings Advanced Signal Processing VI*, 1996, p. 2846.

10. A. Van der Schaaf and J. H. Van Hateren, "Modelling the Power Spectra of Natural Images: Statistics and Information," *Vision Res.*, 1996, pp. 2759–2770

Multidimensional Systems and Signal Processing, 9, 465–473 (1998)
© 1998 Kluwer Academic Publishers, Boston. Manufactured in The Netherlands.

Scale and Translation Invariant Methods for Enhanced Time-Frequency Pattern Recognition

WILLIAM J. WILLIAMS*   wjw@eees.umich.edu   wjw@w.imap.itd.umich.edu
Electrical Engineering and Computer Science Dept., 4230C EECS Bldg., University of Michigan, Ann Arbor MI 48109

EUGENE J. ZALUBAS
Electrical Engineering and Computer Science Dept., 4313 EECS Bldg., University of Michigan, Ann Arbor MI 48109

ROBERT M. NICKEL
Electrical Engineering and Computer Science Dept., 4313 EECS Bldg., University of Michigan, Ann Arbor MI 48109

ALFRED O. HERO III
Electrical Engineering and Computer Science Dept., 42329 EECS Bldg., University of Michigan, Ann Arbor MI 48109

Received October 4, 1996; Revised February 12, 1997

Abstract. Time-frequency (t-f) analysis has clearly reached a certain maturity. One can now often provide striking visual representations of the joint time-frequency energy representation of signals. However, it has been difficult to take advantage of this rich source of information concerning the signal, especially for multidimensional signals. Properly constructed time-frequency distributions enjoy many desirable properties. Attempts to incorporate t-f analysis results into pattern recognition schemes have not been notably successful to date. Aided by Cohen's scale transform one may construct representations from the t-f results which are highly useful in pattern classification. Such methods can produce two dimensional representations which are invariant to time-shift, frequency-shift and scale changes. In addition, two dimensional objects such as images can be represented in a like manner in a four dimensional form. Even so, remaining extraneous variations often defeat the pattern classification approach. This paper presents a method based on noise subspace concepts. The noise subspace enhancement allows one to separate the desired invariant forms from extraneous variations, yielding much improved classification results. Examples from sound classification are discussed.

Key Words: time-frequency, scale, speech, pattern recognition

1. Introduction

Time-frequency (t-f) analysis is useful for signals which exhibit changes in frequency content with time. The well-known spectrogram often presents serious difficulties when it is used to analyze rapidly varying signals, however. If the analysis window is made short enough to capture rapid changes in the signal it becomes impossible to resolve signal components which are close in frequency within the analysis window duration. If the window is made long to resolve the frequencies of sinusoids, the time when sinusoidal components act becomes difficult to determine.

---

*   Supported in part by ONR Grants nos. N00014-90-J-1654 and N000014-97-1-0072, NSF Grant BCS 9110571 and DOD Contract no. MDA904-95-C-2157

Until recently, there was one alternative t-f analysis technique which was widely believed to avoid some of the problems of the spectrogram. The well known Wigner distribution (WD) avoids the problems of windowing and enjoys many useful properties, but often produces an unacceptable amount of interference or cross-term activity between signal components when the signal consists of many components [3,5]. Despite its shortcomings, the Wigner distribution has been employed as an alternative to overcome the resolution shortcomings of the spectrogram. It also provides a high resolution representation in time and in frequency. The WD has many important and interesting properties.

Both the spectrogram and the WD are members of Cohen's Class of Distributions [2]. Cohen has provided a consistent set of definitions for a desirable set of t-f distributions which has been of great value in this area of research. Different members of Cohen's class can be obtained by using different kernels. In this framework, the WD has a unity valued kernel. Choi and Williams introduced the Exponential Distribution (ED), with kernel $\phi_{ED}(\theta, \tau) = e^{-\theta^2\tau^2/\sigma}$, where σ is a kernel parameter ($\sigma > 0$) [1]. The ED overcomes several drawbacks of the spectrogram and WD, providing high resolution with suppressed interferences. A recent comprehensive review by Cohen [3] provides an excellent overview of TFDs and recent results using them.

The Reduced Interference Distribution (RID), which is a generalization of the ED, shares many of the desirable properties of the WD, but also has the important reduced interference property. RID is discussed in a recent book chapter [13] and a design procedure for RID kernels has been developed [7]. One may start with a primitive function, $h(t)$, which has certain simple constraints, and evolve a full-fledged RID kernel from it. The RID kernel retains a unity value along the θ and τ axes in the ambiguity plane, generally providing good time-frequency resolution and auto-term preservation, but attenuates strongly elsewhere for good cross-term suppression.

1.1. The Scale Transform

The scale transform has been described by Cohen [4] to be:

$$D(c) = \frac{1}{\sqrt{2\pi}} \int_0^\infty x(t) \frac{e^{-jc\ln t}}{\sqrt{t}} dt \tag{1}$$

The scale transform has an analogy to the Fourier transform. The Fourier transform of a signal, $x(t)$ and the Fourier transform of a shifted version of that signal, $x(t - t_o)$ differ only by a phase factor.

$$F[x(t - t_o)] = X_o(\omega) = X(\omega)e^{-j\omega t_o} \tag{2}$$

so that

$$|X(\omega)| = |X_o(\omega)|. \tag{3}$$

In a like manner, the scale transform of $\sqrt{|a|}x(at)$ differs from the scale transform of $x(t)$

only by a phase factor, so that the magnitudes of the scale transform of $x(t)$ and $\sqrt{|a|}x(at)$ are identical.

$$|D(c)| = |D_a(c)| \tag{4}$$

We have developed discrete forms of the scale transform [14,16] which can be computed efficiently. One might question the use of the scale transform rather than the more well-known Mellin transform. There are several reasons for using the scale transform. One reason is that the standard Mellin transform weights signal components in lower time more than in higher time. A second reason is the relationship of scale to wavelet concepts and the insights it brings in this light.

2. Acoustic Signals

Two types of acoustic signals were used to test the effectiveness of these methods. These were human speech and marine mammal sounds.

2.1. Marine Mammal Sounds

Marine mammal sounds are well characterized using the RID and overcome some of the shortcomings of the SP as described by Watkins [12]. RID clearly reveals both the tonal structure in the whistles and the temporal structure of clicks which are simultaneously produced by these animals. It appears that the clicks of marine mammals such as whales and dolphins may have a distinctive structure based on the individual animal and may be useful in nonintrusive tagging and tracking of these animals. Our new TFD methods provide a powerful means of representing the complex sounds produced by marine mammals.

One can now readily design TFDs which represent the joint energy of a signal as a function of time and frequency or space-frequency distributions which represent the joint energy of images as space-spatial frequency distributions (two spatial variables x and y and two spatial frequency variables Ω_X and Ω_Y). Furthermore, with careful design, these joint distributions can exhibit proper covariances with time, frequency or spatial shifts such that the representation shifts in accordance with these shifts but does not change in its configuration [14]. The well-known spectrogram has been extensively used in speech analysis and it has these useful properties. A shift in time or a shift in frequency of the signal[1] will shift the representation appropriately in time and frequency. However, the spectrogram does not exhibit the proper characteristics in response to *scale* changes in the signal. That is if $x(t)$ becomes $x(at)$, the Fourier transform of $x(t)$ changes from $X(\omega)$ to $\frac{1}{a}X(\frac{\omega}{a})$. This is illustrated in Figure 1.

2.2. Speech Processing

The spectrogram has long been a widely used tool in speech analysis. Other TFDs have been investigated in speech analysis, but none have yet provided a strong advance beyond

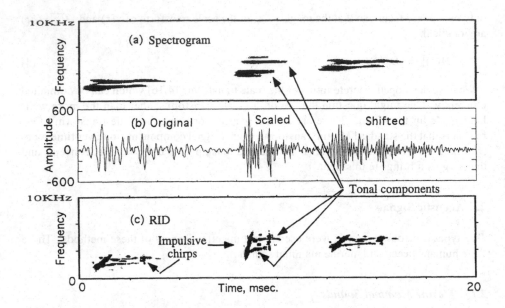

Figure 1. TFD results for time shifted, frequency shifted and scaled dolphin click. a. Spectrogram. b. Original click, scaled and time shifted click, time shifted and frequency shifted click. c. Reduced Interference Distribution (RID) click results for the same time shifts, scaling and frequency shift. (From: W. J. Williams, Reduced Interference Distributions: Biological Applications and Interpretations, *IEEE Proc.*, Vol 84, 1996, pp 1264–1280.)

the level of analysis provided by the spectrogram [9]. It is generally believed that the potential exists for significant advances in speech analysis using recently developed TFD tools, however.

3.   Classification and Detection of Acoustic Signals

Acoustic signals may vary in time of occurrence, frequency and scale. The Doppler effect manifests itself as scaling. The "scalogram" [10], which is an analog of the spectrogram with frequency replaced by scale, might be effective in analyzing scaled signals. However, it, as does the wavelet transform, lacks the frequency covariance property. One might like to have invariant representations under time-shift, frequency-shift and scale. The techniques described in this paper may be assembled to achieve all of these invariances. First, the sound or a segment of the sound to be analyzed must be isolated. Next, the RID is computed. Then, the autocorrelation[2] along time is performed as

$$A_{RID}(m, k) = \sum_n RID(n, k) RID(n - m, k) \qquad (5)$$

where n is the time sample, k is the frequency sample and m is the autocorrelation lag sample. This removes absolute time and produces a centered autocorrelation of each frequency slice.

Next, one has the choice of also removing absolute frequency by performing a similar autocorrelation along k. Finally, one may scale transform the resulting representation along the time and frequency directions. This serves to produce a representation that is invariant to time, frequency and scale shifts.

It may not be desirable to remove all variation. These variations may serve to classify or detect the signal. For example, if frequency shift is an important indicator of the identity of a signal, one may bypass that step. It is important to note that even though these techniques serve to make the representations invariant, true frequency shifts and scale shifts may be retained in the phases of the requisite transforms.

3.1. Classification and Detection of Images

Recognizing characters or spotting words in bitmapped documents (images) has been of particular interest to us [14]. One may convert an image into a four dimensional representation in a manner analogous to the conversion of a one dimensional signal into a t-f representation. Here, the two spatial dimensions (x, y) are retained and joined by the spatial frequencies (ω_x, ω_y). Four dimensional kernels may be applied in a manner similar to t-f analysis. We have developed software which accomplishes this and it is clear that these 4-D representations are rich in detail. The 4-D structure does not prevent the application of the invariance producing transformations along all four dimensions, however. In order to reduce the complexity of the representations, we have reduced the image computations to two dimensional autocorrelations along x and y with the idea of expanding to the full four dimensional forms after working out details in the simplified formulations. Noise subspace and higher order statistical moments have been gainfully applied to this problem [6].

3.2. Applying the Scale Transform

One of the problems in applying the scale transform is finding the zero reference. Unlike the Fourier transform, the scale transform exhibits strong non-stationary characteristics. However, the process of autocorrelation provides an unequivocal zero reference for time. Likewise, the frequency dimension of the RID has an unequivocal zero frequency reference, so that the scale transform may be easily applied. Due to symmetries, the 2-D scale transform may only need to be applied to unique quadrants of the autocorrelated RID representation or the 2-D autocorrelated images. Previous results in using the 2-D scale transform to render 2-D autocorrelated images invariant to displacement and scale are very encouraging [14].

3.3. Sound Classification using the Invariant Representations

Starting with a suitable TFD[3], almost all of the undesired variation due to time shift, frequency shift and scale may be squeezed out of the final invariant form. There may still be some residual effects due to discretization and computation. The next task is to design a classifier. Suppose that the invariant form is characterized by a two dimensional representation $\Delta(p, q)$. This 2-D representation may be decomposed using eigensystem techniques as

$$\Delta(p, q) = \sum_j a_j \beta_j(p, q) \tag{6}$$

where the $\beta_j(p, q)$ are eigenimages and the a_j are the eigenvalues of the decomposition. The eigensystem decomposition is carried out on a collection of $\Delta(p, q)$ examples coming from the classes of objects (signals or images) that are of interest. The eigensystem decomposition then provides an ordered set of eigenimages ordered according to their eigenvalues. Although the eventual goal is to use true two dimensional eigenimage analysis, suitable algorithms to accomplish this have not been identified. One may utilize a simpler one dimensional approach which lends itself to readily available algorithms.

The 2-D $N \times M$ invariant forms may be converted into vectors of length $N \times M$ by either concatenating the rows or columns. Then, readily available Singular Value Decomposition (SVD) techniques may be applied to the vectorized set of images. Suppose there are several different extraneous variations in the supposedly invariant representations caused by a variety of factors. For example, the same person may not say the same word exactly the same way each time or the same whale or dolphin may click slightly differently each time. Such extraneous variations often confound the invariant representations so that effective detection or classification of a specific signal or image is rendered impossible. A new and very effective method using noise subspace concepts has been developed to overcome these problems.

4. Noise Subspace Methods

The $N \times M$ vectorized 2-D forms have a large number of elements. Usually, for classification methods to work, one wishes to have a considerably greater number of representations of the signal vectors than there are elements in those representations. Here, we have exactly the opposite. There are many more elements in the vectorized 2-D forms than there are vectorized 2-D forms. This is usually a statistical nightmare. However, suppose there are K sound examples ($K \ll N \times M$). Then the SVD produces $N \times M$ orthogonal eigenvectors, the first K of which form a complete orthonormal basis for the vectorized invariant forms. The remaining SVD eigenvectors (the noise eigenvectors) must be orthogonal to all of the original vectorized invariant forms. Suppose that we now obtain a new signal. Convert it into the TFD, then to the 2-D invariant form and finally, vectorize the 2-D invariant form. If it belongs to the set of vectorized 2-D invariant forms used to produce the SVD results, then it should be **orthogonal** to all of the noise eigenvectors produced by the SVD. Therefore, its projection on any of the noise eigenvectors should be zero. If we have carried out the whole process through the SVD for a number of different sets of signals, we should find the projection of the vectorized 2-D invariant form of the unknown signal on the noise eigenvectors of each set of signals. The smallest result will be theoretically obtained when this is done using the noise eigenvectors of the set to which the signal belongs.

5. Sound Classification Results

Two different experiments were carried out. Invariant forms as described were derived from the signals. In one experiment the clicks from two sperm whales were considered. In the second experiment, speaker identification was attempted. Ten speakers said 'Michigan' ten times. Half the data sets were used to develop the classifier and the technique was tested using the remaining data sets. In both experiments, classification success of individuals was quite high. Some results are shown in Figure 2. Half of the responses for each individual were used to develop a classifier and the other half to test it. Only autocorrelation along time was applied to the RID result in the human speaker study. It is believed that variations in the frequency direction should be retained, since they may serve to identify individual speakers. The methodology reported in this paper is one of two new techniques we have developed recently. The other (moment-based) method [17] is also quite effective in sound

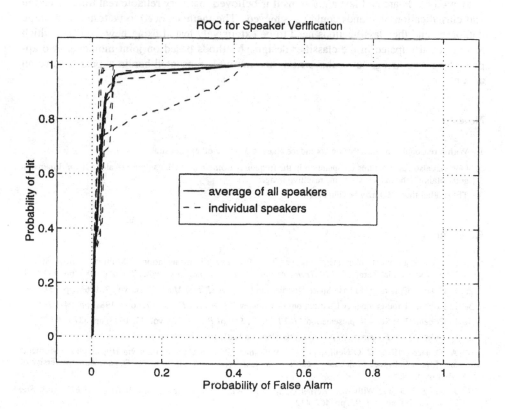

Figure 2. Individual ROC results for ten speakers and the average ROC for all speakers.

pattern recognition. Tacer and Loughlin [11] have had success in untilizing joint TFD moments in classifying postural sway.

6. Conclusions

The methods outlined in this paper seem to work very well. At this point the ideas have been applied in a straightforward way with very little "tuning" of the various components of the technique. At this time the approach appears to be competitive, at least, with alternative methods of speaker identification where complex and highly developed alignment algorithms and time warping algorithms have been applied together with commonly used pattern recognition engines. Further refinements of our technique may provide significant improvements over present results. It is believed the technique could be applied to a wide variety of sounds, other signals and images in terms of specific identification of distinct classes of signals and images. There are a number of trade-offs to be considered and we have a number of improvements in mind. The computational burden is high, but with dedicated hardware and fast algorithms it is believed that very reliable real time detection and classification of sounds could be achieved. The methods need to be tested with large databases and the classifications need to be extensively tested using new data sets which have not participated in the classifier design. Methods based on joint moments also appear to be promising, but at this writing the methods presented herein appear to have an advantage.

Notes

1. Within reasonable bounds that do not induce aliasing or some other undesirable effect.
2. One can also carry out this computation in the frequency domain as well, using FFTs along the n dimension and obtaining the magnitude of the resulting image.
3. TFDs other than RID may be suitable [8].

References

1. H. I. Choi, and W. J. Williams, "Improved Time-Frequency Representation of Multi-component Signals Using Exponential Kernels," *IEEE Trans. Acoust., Speech, Signal Proc.*, vol. 37, no. 6, 1989, pp. 862–871.
2. L. Cohen, "Generalized Phase-Space Distribution Functions," *J. of Math. Phys.*, vol. 7, 1966, pp. 781–786.
3. L. Cohen, "Time-Frequency Distributions—A Review," *Proc. IEEE*, vol. 77, no. 7, 1989, pp. 941–981.
4. L. Cohen, "The Scale Representation," *IEEE Trans. Signal Processing*, vol. 41, 1993, pp. 3275–3292.
5. L. Cohen, *Time-Frequency Signal Analysis*, Prentice Hall, 1995.
6. A. O. Hero, III, J. C. O'Neill, and W. J. Williams, "Pattern Classification via Higher Order Moments and Signal Subspace Projections," CSPL Technical Report No. 20-96 EECS Department, University of Michigan, 1996.
7. J. Jeong W. J. and Williams, "Kernel Design for Reduced Interference Distributions," *IEEE Trans. Sig. Proc.*, vol. 40, no. 2, 1992, pp. 402–412.
8. P. J. Loughlin, "Comments on Scale Invariance of Time-Frequency Distributions," *IEEE-Signal Processing Letters.*, vol. 2, 1995, pp. 4–6.

9. J. W. Pitton, K. Wang, and B.-H. Juang, "Time-Frequency Analysis and Auditory Modeling for Automatic Recogition of Speech," *IEEE Proc.*, vol. 84, 1996, pp. 1199–1215.

10. O. Rioul, and P. Flandrin, "Time-Scale Energy Distributions: A General Class Extending Wavelet Transforms," *IEEE Trans. Sig. Proc.*, vol. 40, no. 7, 1992, pp. 1746–1757.

11. B. Tacer, and P. J. Loughlin, "Time-frequency Based Classification," *Advanced Signal Processing Algorithms, Architectures and Implimentions, VI*, SPIE vol. 2846, 1996, pp. 186–192.

12. W. A. Watkins, "The Harmonic Interval: Fact or Artefact in Spectral Analysis of Pulse Trains," *Marine Bio-acoustics*, vol. 2, 1966, pp. 15–43.

13. W. J. Williams, and J. Jeong, "Reduced Interference Time-Frequency Distributions," B. Boashash Ed., *Time-Frequency Signal Analysis*, Longman and Cheshire—J. W. Wiley, 1992.

14. W. J. Williams, E. J. Zalubas, and A. O. Hero, III, "Separating Desired Image and Signal Invariant Components from Extraneous Variations," *Advanced Signal Processing Algorithms, Architectures, and Implementations, SPIE Proceedings*, vol. 2846, 1996

15. W. J. Williams, "Reduced Interference Distributions: Biological Applications and Interpretations," *IEEE Proc.*, vol. 84, 1996, pp. 1264–1280.

16. E. J. Zalubas, and W. J. Williams, "Discrete Scale Transform for Signal Analysis," *Proc. IEEE Intl. Conf. on Acoustics, Speech and Signal Processing*, vol. 3, 1995, pp. 1557–1561.

17. E. J. Zalubas, J. C. O'Neill, W. J. Williams, and A. O. Hero III, "Shift and Scale Invariant Detection," *IEEE Int. Conf. Acous., Speech, Sig. Proc.*, vol. 5, 1997, pp. 3637–3640.